T0267389

Managing Change in Construction Projects
A Knowledge-Based Approach

Managing Change in Construction Projects
A Knowledge-Based Approach

Sepani Senaratne

Department of Building Economics,
University of Moratuwa, Sri Lanka

&

Martin Sexton

School of Construction Management and Engineering
University of Reading, UK

A John Wiley & Sons, Ltd., Publication

Blackwell Publishing was acquired by John Wiley & Sons in February 2007. Blackwell's publishing programme has been merged with Wiley's global Scientific, Technical, and Medical business to form Wiley-Blackwell.

Registered Office
John Wiley & Sons Ltd, The Atrium, Southern Gate, Chichester, West Sussex, PO19 8SQ, UK

Editorial Offices
9600 Garsington Road, Oxford, OX4 2DQ, UK
The Atrium, Southern Gate, Chichester, West Sussex, PO19 8SQ, UK
2121 State Avenue, Ames, Iowa 50014-8300, USA

For details of our global editorial offices, for customer services and for information about how to apply for permission to reuse the copyright material in this book please see our website at www.wiley.com/wiley-blackwell.

Library of Congress Cataloging-in-Publication Data

Senaratne, Sepani.
  Managing change in construction projects : a knowledge-based approach / Sepani Senaratne & Martin Sexton.
    p. cm. – (Innovation in the built environment)
  Includes index.
  ISBN 978-1-4443-3515-6
  1. Building–Superintendence. 2. Building–Information services. I. Sexton, Martin, 1966-II. Title.
  TH438.S445 2011
  690.068–dc22
                                                    2010030546

A catalogue record for this book is available from the British Library.

This book is published in the following electronic formats: ePDF (9781444392401); ePub (9781444392418)

Set in 10/12 Sabon by SPi Publisher Services, Pondicherry, India

1 2011

Innovation in the Built Environment

Series advisors

Carolyn Hayles, *Queen's University, Belfast*
Richard Kirkham, *University of Manchester*
Andrew Knight, *Nottingham Trent University*
Stephen Pryke, *University College London*
Steve Rowlinson, *The University of Hong Kong*
Derek Thompson, *Heriot Watt University*
Sara Wilkinson, *University of Melbourne*

Innovation in the Built Environment (IBE) is a new book series for the construction industry published jointly by the Royal Institute of Chartered Surveyors and Wiley-Blackwell. It addresses issues of current research and practitioner relevance and takes an international perspective, drawing from research applications and case studies worldwide.

- presents the latest thinking on the processes that influence the design, construction and management of the built environment

- based on strong theoretical concepts and draws on both established techniques for analysing the processes that shape the built environment – and on those from other disciplines

- embraces a comparative approach, allowing best practice to be put forward

- demonstrates the contribution that effective management of built environment processes can make

Published and forthcoming books in the IBE series

Akintoye & Beck, *Policy, Finance & Management for Public-Private Partnerships*
Lu & Sexton, *Innovation in Small Professional Practices in the Built Environment*
Boussabaine, *Risk Pricing Strategies for Public-Private Partnerships*
Kirkham & Boussabaine, *Whole Life-Cycle Costing*
Booth, Hammond, Lamond & Proverbs, *Solutions to Climate Change Challenges in the Built Environment*
Pryke, *Construction Supply Chain Management: Concepts and Case Studies*

We welcome proposals for new, high quality, research-based books which are academically rigorous and informed by the latest thinking; please contact Stephen Brown or Madeleine Metcalfe.

Stephen Brown
Head of Research
RICS
12 Great George Street
London SW1P 3AD
sbrown@rics.org

Madeleine Metcalfe
Senior Commissioning Editor
Wiley-Blackwell
9600 Garsington Road
Oxford OX4 2DQ
mmetcalfe@wiley.com

Contents

About the Authors

Dr Sepani Senaratne, BSc (Hons) in QS, PhD

Sepani Senaratne is a senior lecturer at the department of Building Economics, University of Moratuwa, Sri Lanka. She is the Research Director of the department's research unit – Building Economics and Management Research Unit (BEMRU) and the main contact from the department for the institutional membership of the International Council for Research and Innovation in Building and Construction (CIB). Her main research area includes knowledge management within construction project settings, construction project teams and their learning processes. She has published nearly 40 journal and conference papers and presented papers at several international forums.

Professor Martin Sexton, BSc, MSc, PhD

Martin Sexton is a professor in Construction Management and Innovation at the University of Reading. His research interests range across the organization and management of construction – with a particular focus on understanding the nature and process of innovation at sector, company and project levels. Martin is the Joint Co-ordinator of the International Council for Research and Innovation in Building and Construction (CIB) Working Commission 65 in the Organisation and Management of Construction. He has published widely including 3 books and over 150 journal and conference papers.

Introduction

Construction projects often experience delays, cost overruns and non-conformance to quality, all of which lead to poor performance and dissatisfied parties (see, e.g., Latham, 1994; Egan, 1998). Egan (1998, p. 8), for example, laments that 'more than a third of major clients are dissatisfied with contractors' performance in keeping to the quoted price and to time, resolving defects, and delivering a final product of the required quality. ... [and] ... more than a third of major clients are dissatisfied with consultants' performance in coordinating teams, in design and innovation, in providing a speedy and reliable service and in providing value for money.' Hence an understanding of the driving forces behind such problems is a priority if ways are to be developed to overcome them and to improve the performance of the industry.

Unexpected change occurring within the design and construction phases can hinder project success to a significant extent (CII-Construction Industry Institute, 1994; CIRIA, 2001). CIRIA (2001, p. 10) defines construction project change as 'an alteration or a modification to the pre-existing conditions, assumptions or requirements.' These project changes are the additions, deletions or revisions within the scope of a project contract that cause an adjustment to contract price, contract time (CII, 1994) and/or quality.

Changes of this nature can directly result in a number of problems within construction projects, especially time overruns, cost overruns and quality deviations. The major cost due to change is the cost of rework, which can amount to 10–15% of the contract value (Love and Li, 2000). Further, the indirect effects of change can be considerable. Examples of indirect effects are loss of productivity, and interruption to workflows and cash flows. These effects, in turn, may lead to low morale and increased claims and disputes between project parties (Bower, 2000). The appropriate management of change is thus essential to the minimisation of the disruptive effects of unexpected change in construction projects. Effective change management allows change to take

Managing Change in Construction Projects: A Knowledge-Based Approach, First Edition.
Sepani Senaratne and Martin Sexton.
© 2011 by Sepani Senaratne and Martin Sexton. Published 2011 by Blackwell Publishing Ltd.

place in a more controlled way so that viable alternatives are identified and developed, and the impact is better defined before implementation.

The impact of change increases when the project moves from the design to the construction stage (CIRIA, 2001). Lawson (1997), when explaining the design process, states that many components of the design problem do not emerge until some attempt is being made to generate solutions. Design is thus full of uncertainties, both about objectives and their priorities. Even though later changes can be reduced significantly by proactive change management approaches, they are still common in the construction phase due to these uncertainties and incomplete project information. The later a change is introduced, the more disruption and cost will be incurred.

Project change can be triggered from internal or external causes. The external causes can emerge from wider business sector and inter-organisational level forces. However, many causes of project changes are internal and are generated from the intra-project environment, which can be either design- or construction-generated (Love et al., 1999). The root of these project-level causes (such as errors, omissions and poor information) can be traced back to the skills and the knowledge of project participants and how these are combined to manage construction projects (Tombesi, 2000). As Manavazhi (2004) notes, construction practitioners assess change and rework largely based on intuition sharpened by many years of experience. Hence, change management needs to address these key issues that trigger project changes.

In construction projects, problem-solving often takes place in team environments (Anumba et al., 2001; Gunasekaran and Love, 1998). The manner, therefore, in which the construction project team-members are integrated and co-ordinated is important (Cox and Townsend, 1997). For example, according to Constructing Excellence (2004, p. 4) 'construction is a collaborative activity – only by pooling the knowledge and experience of many people can buildings meet the needs of today, let alone tomorrow. But simply bringing people together does not necessarily ensure they will function effectively as a team.' Construction participants (e.g. clients, architects, engineers, quantity surveyors, contractors and suppliers) are required to work effectively as a team to deliver projects successfully. This requirement has significant implications for the management of project change: effective project change management does not rely solely on the role of a project manager; rather, it requires appropriate engagement from all relevant team members. Studies in construction reveal that teamwork is significantly determined by contractual arrangements (see, e.g., Cornick and Mather, 1999). The emergence of collaborative practice in the construction industry, which includes new team approaches such as design and build (D&B), project management, partnering and other partnership arrangements provide opportunities to improve teamworking (Muir and Rance, 1995). There is a dearth of literature, however, on the management of project change within such collaborative team settings.

In summary, the effective management of change is a necessity to minimise the disruptive effects of change in construction projects. Change during the construction phase is more disruptive and costly than change during the design phase. There is a deficiency of research on the management of change during the construction phase in collaborative team environments. The literature review (see Chapter 2) identifies that managing team knowledge is of significant importance for the effective management of project change. However, the problem with this literature is its emphasis on introducing various tools and techniques to systemise the change process without properly understanding the key roles that the knowledge and experience of the participants play in managing projects. The need for this understanding is the point of departure for this book. The next section justifies this argument while briefly explaining the conceptual issues and gaps within the literature.

1.2   Need for the Investigation

Previous approaches to construction project change management adopt a variety of different perspectives. CII (1994) and CIRIA (2001), for example, take a general change management perspective by providing best practice guidelines on project change management. These guidelines are based on five principles: anticipate change; recognise change; evaluate change; resolve change; and learn from change. These principles aim to mitigate the disruptive effects of changes by suggesting a change management framework established at the start of the project. Love *et al.* (1999) take a more technical perspective by addressing the rework effects of project change. Their work confirms the complexity and the interdependence of project changes with the identification of various causes and effects of project changes. Other studies have approached project change from a process management perspective. Kagioglou *et al.* (2000), for example, introduce a separate change management process within the generic design and construction process protocol.

Drawing from the previous construction project change literature, it is evident that problem-solving has been viewed essentially as an information-processing activity rather than a knowledge-intensive activity (see, e.g., Winch, 2002). The information-processing perspective on organisation originates from the work of Simon (1957) and Galbraith (1974), which asserts that the key feature of organisation is to process information to enable managers to make better decisions. The assumption underpinning this perspective is that organisations should match their information-processing activities to their information needs (see, e.g., Daft and Lengal, 1986). However, empirical research has found that information processing across organisation boundaries presents significant barriers to effectiveness. Successful project delivery requires the development and application of a wide range of specialist knowledge located in different actors, and that actors mutually 'know' how their roles fit with each other. This cognitive dimension cannot be overcome by information processing alone. The limitation of the information-processing view has stimulated the development of an alternative theory of the firm, which recognises that 'knowledge is the key asset' and 'knowing is the key process', in delivering organisations' competitive advantage. On the 'knowledge as an asset' perspective knowledge is often viewed as an objectively definable commodity, which can be managed and controlled by certain mechanisms. For 'knowing as a process' views, knowledge is a social construct, developed, transmitted and maintained in social situations. These knowledge-based views of the firm (Grant, 1996a, 1996b; Spender, 1996; Empson, 2001) open new avenues to approach effective project change management in construction.

The construction literature that addresses knowledge management, learning and innovation, shows a trend towards identifying construction problem-solving as a knowledge-intensive activity. Egbu *et al.* (2003), for example, identify project problem situations as a key trigger of knowledge production in construction. Winch (2002) explains that knowledge and learning are generated in solving problems that involve team discussions and dialogues during the construction process. For such problem-solving to become true innovation, the solutions reached for particular problems should be learned, codified and applied in future projects (Sexton and Barrett, 2003; Lu and Sexton, 2009). Similarly, other learning and innovation literature in construction identifies the importance of integrating project experience to the organisational business processes, to generate learning and innovation (see, e.g., Barlow and Jashapara, 1998; Gann and Salter, 2000). However, the extant knowledge-based construction literature arguably does not provide an in-depth understanding of the role of knowledge during construction problem-solving, especially during managing change.

To this end, the general knowledge management literature aids in understanding the fuller role of knowledge during problem situations that is facilitated by team interactions. Accordingly, during shared activities, such as problem-solving, individuals bring various forms of knowledge that could be shared and converted into new knowledge (Nonaka and Takeuchi, 1995; Leonard and Sensiper, 1998). When considering team knowledge during change events, the theory of knowledge creation (Nonaka and Takeuchi, 1995) shows how a team can advance knowledge and learning through team interactions. However, as Snowden (2002) argues, tacit knowledge need not necessarily go through a costly codification process to create new knowledge. This understanding on knowledge creation offers significant contributions in understanding the role of knowledge during shared activities. In order for knowledge to be useful it needs to pass from project to organisation level, and then back to parallel and subsequent projects. This inter-project learning can emerge when team knowledge is stored and transferred within the organisation for re-use in future projects.

In summary, this book identifies the problem of construction projects with disruptive effects due to unplanned changes. Previous approaches to managing project change adopt an information-processing view, without appreciating the significant role of knowledge in managing change. This led to the exploration of the role of knowledge during team interaction as explained in the knowledge management literature. From this knowledge-based perspective of managing project change, the research problem is articulated below.

1.3   Questions to Research

The aim of this book is to explore the role of knowledge during the 'management of unplanned change in the construction phase within collaborative team settings' (referred to in this research as 'reactive change process'). This aim is progressively developed through the following research questions:

> RQ1: *What are the key contextual factors of the reactive change process?*
> RQ2: *What are the properties of knowledge that the project team members use in the reactive change process?*
> RQ3: *How does the project team identify and utilise this knowledge during the reactive change process?*
> RQ4: *How does the knowledge that the project team use in the reactive change process, interact and form new knowledge?*
> RQ5: *How is the knowledge that is created through the reactive change process, transferred and disseminated within the multiple organisations for potential re-use in future projects?*

1.4   Summary and Link

This chapter has set out the background and principal focus for this book. The next chapter will contextualise the outlined research issues within the relevant general and construction-specific change management literature.

Key Issues from the Literature

2.1 Introduction

Chapter 1 introduced the principal purpose of this book. The aim of this chapter is to identify key issues in the literature in order to develop a clear understanding of the research problem. The chapter is structured as follows:

- First, the construction project environment is explained in order to understand the broad context within which projects are located. The discussion starts by explaining the unique characteristics of construction. Then, the construction project team is described. The emerging collaborative team approach in construction is subsequently discussed. Finally, teamwork issues in construction settings are articulated (see Section 2.2).
- Second, the core research area of this book, construction project change management, is addressed. The discussion starts with the identification of general change management issues before moving on to construction project change, which is explained in terms of its nature, process, causes, effects, context and approaches (see Section 2.3).
- Third, to understand project change process as a knowledge-processing activity, the chapter introduces a knowledge management perspective. This is approached by explaining the evolution of knowledge-based theories and knowledge-based views in the construction literature (see Section 2.4).
- Fourth, the role of knowledge during construction project change is described in terms of properties of knowledge, knowledge creation, knowledge transfer and their influencing factors (see Section 2.5).
- Finally, the research problem statement is set out by integrating the arguments from the literature to offer a knowledge-based project change process (see Section 2.6).

Managing Change in Construction Projects: A Knowledge-Based Approach, First Edition.
Sepani Senaratne and Martin Sexton.
© 2011 by Sepani Senaratne and Martin Sexton. Published 2011 by Blackwell Publishing Ltd.

2.2   Construction Project Environment

2.2.1   Nature of construction

The construction project environment is generally viewed as being distinctive from other sectors due to a number of unique characteristics. These characteristics can be grouped into six areas. First, the construction product is characterised by factors such as immobility, complexity, durability, low technology and high cost (Nam and Tatum, 1988), which makes it different and more complex compared with other products; for example, high-technology manufacturing products.

Second, a construction project is generally custom-built to a client's bespoke needs, making it unique and one-off (Cornick and Mather, 1999; Gann and Salter, 2000). Construction projects, even though they are often unique in the specific combination of client, supplier firms, location, time and the social, economic and political environment, can be generic in terms of particular processes. For example, the generic design and construction process protocol (Kagioglou *et al.*, 2000) provides a range of processes that are generic to a construction project. Thus, despite the unique nature of construction projects, similar processes provide scope to deliver projects in a consistent way and to learn from one project to another.

Third, construction projects are generally delivered by temporary organisations, where a new team is formed for each building. The implication of this transitory nature is that 'the management of innovation is complicated by the discontinuous nature of project-based production processes, in which there are often broken learning and feedback loops' (Gann and Salter, 1998, p. 435). Project-based firms thus often lack the organisational mechanisms for the knowledge acquired in one project to be transferred and used by other projects (Prencipe and Tell, 2001; Lu and Sexton, 2009). However, emergence of collaborative arrangements such as partnering (discussed in Section 2.2.3) may provide long-term learning curves within construction (Black *et al.*, 2000).

Fourth, a construction project consists of team members who come from different backgrounds and cultures, to form a 'multiple organisation' (Cornick and Mather, 1999). Therefore, construction project settings not only differ from permanent settings but also from other project-based settings where teams are formed by drawing individuals from a single company. In construction, as Bresnen *et al.* (2003) point out, discontinuities are added by the fragmentation of the construction project team into different professional disciplines. A construction project team, for example, can be made up of architects, contractors, suppliers and financiers. The differences between parties encourage them to prioritise their own interests to the detriment of the overall project interest and, in so doing, compete against each other. The construction industry, as a consequence, is often characterised by its deep-rooted adversarial culture (Gil *et al.*, 2001), which works against successful teamworking.

Fifth, the construction industry is fragmented, with a large number of firms that are geographically distributed (Latham, 1994). The project team frequently becomes a virtual team, as it has to work together from many different locations, over the life of a project. Beyond the immediate project team, construction firms need to further manage networks of suppliers, customers and regulatory bodies, as knowledge is differentiated and distributed within, and across, these supply networks (Gann and Salter, 1998).

Sixth, there is a growing body of knowledge that emphasises the knowledge-intensive nature of construction. Lu and Sexton (2009) note the knowledge intensification of construction in line with the broad trend towards knowledge economies. This intensification, they argue, is indicated by the sharp and ongoing rise in the scale and scope of knowledge-intensive professional service firms.

On the whole, the long-established nature of construction, with characteristics such as complex products, one-off nature, temporary multiple organisational settings, adversarial cultures and fragmentation, can significantly hinder learning and knowledge advancement within the industry (the issue of learning and knowledge dissemination in construction is addressed further in Section 2.4.2). At a project level, knowledge and learning are very much affected by the activities of project team members.

The manner in which the construction project team members are integrated and co-ordinated is, therefore, important (Cox and Townsend, 1997). According to Constructing Excellence (2004, p. 4), for example, 'construction is a collaborative activity – only by pooling the knowledge and experience of many people can buildings meet the needs of today, let alone tomorrow. But simply bringing people together does not necessarily ensure they will function effectively as a team.' Construction participants (e.g. clients, architects, engineers, quantity surveyors, contractors and suppliers) are required to work effectively as a team to deliver projects successfully. Accordingly, understanding the nature of the construction team and the issues that can result from construction teamworking is important and is discussed in the next section.

2.2.2   Construction project team

A review of the general literature on teams and groups reveals a lack of consensus on a single definition of a 'team' or a 'group'. According to Fishers *et al.* (1997), there are two categories of authors: those who use the terms 'team' and 'group' interchangeably and imply that the 'work team' is similar to the 'work group'; and those who view the 'team' as a 'group', but with something extra. This distinction between group and team can be explained through the following definitions. A 'group' is any number of individuals who interact with one another, are psychologically aware of one another, and perceive themselves to be a group (Schein 1970, cited in Fisher *et al.*, 1997). Conversely, a 'team' is a small number of people with complementary skills who are committed to a common purpose, performance goals and approach for which they hold themselves mutually accountable (Katzenbach and Smith, 1993a, 1993b). Thus, a 'team' is essentially a 'group', but with additional characteristics such as complementary skills, mutual accountability and commitment.

In construction, since members from different backgrounds with complementary skills need to work together towards the project goals and are accountable for their individual tasks, they can be regarded as a team, as opposed to a group. Cornick and Mather (1999) compare and contrast a construction team with an ideal team (such as a sports team): although the construction team is committed to a common purpose, which is determined by the client, they differ from an ideal team, as they have a vested interest in their own firm, which may or not be the same as the overall project interest. Further, in construction the team composition is generally determined by technical and financial considerations rather than teamworking abilities. Above all, the method of working in construction projects is determined by contractual conditions, which might promote or hinder a committed approach and mutual accountability. A recent study confirms that construction project teams have these features, with the following definition: 'the construction team is a collection of two or more people with complementary skills, who come from different disciplines and organizations, to perform a common objective, but with individual objectives and, operating from different locations with multiple reporting relationships, whose accountability and leadership are significantly governed by the contractual arrangements' (Senaratne and Hupuarachchi, 2009, p. 223). With this emphasis on contractual conditions in construction settings, the next section explores procurement arrangements and their relationship to teamworking.

2.2.3   Movement towards collaborative team approaches

As Constructing Excellence (2004) proposes, to improve construction teamwork the projects need to move away from the traditional adversarial culture and build a team culture that includes shared vision, joint responsibilities, openness, mutual trust and regular reviews. Muir and Rance (1995) explain the emergence of collaborative practice in the construction industry, which includes new team approaches such as design and build, project management, partnering and other partnership arrangements.

Among these new team approaches, partnering is becoming popular as a means of achieving 'collaborative' and 'teamwork' approaches to solve construction industry problems (see, e.g., Black *et al.*, 2000; Bresnen and Marshall, 2000; Humphreys *et al.*, 2003). Partnering is defined as 'a management approach used by two or more organisations to achieve specific business objectives by maximising the effectiveness of each participant's resources. The approach is based on mutual objectives, an agreed method for problem resolution, and an active search for continuous measurable improvements' (Reading Construction Forum, 1995, p. 2). Partnering can be applied with most procurement methods in use, and can be either project partnering used in one-off projects or strategic partnering used as long-term arrangements between clients, contractors, consultants, and even with the subcontractors and suppliers down the supply chain. Partnering arrangements provide opportunities for long-term team learning and effective sharing of knowledge between the parties.

Other integrative or collaborative approaches, as Baiden *et al.* (2006) identify, include design-and-build procurement paths. In these arrangements, contractors and consultants form a single organisation, known as a design-and-build firm. Due to features such as single point responsibility, improved project performance and better client satisfaction, this method has become popular in the UK construction industry (Akintoye, 1994). The design-and-build integrated approach not only provides opportunities to leverage speciality knowledge of contractors but also enables joint creation of new knowledge (Gil *et al.*, 2001).

On the whole, the above-explained team approaches in construction have been introduced to overcome the adversarial culture in construction (Latham, 1994). However, empirical studies repeatedly show that the adversarial culture itself has been the barrier to gaining effective benefits from such approaches. Even though the collaborative approaches provide opportunities to increase team interaction, in order to gain real benefits of collaborative arrangements, an open team environment against the traditional adversarial environment needs to be explicitly established (Moore and Dainty, 1999, 2001; Bresnen and Marshall, 2000; Gil *et al.*, 2001). Baiden *et al.* (2006) use the term 'integrated construction project team' to describe this type of collaboration in construction projects. Hence, to achieve an integrated construction project team, effective teamworking is a recurring issue despite the emergence of collaborative arrangements in construction. The next section captures these teamwork issues related to construction.

2.2.4   Construction teamwork issues

Teamwork in construction starts with the forming of a project team (Busseri and Palmer, 2000). From this point, as Adair (1986) explains, the team faces issues such as defining project goals against individual goals and clarifying team roles including leadership and client roles. The project-based nature of construction projects disrupts the team development process. Although the construction team undergoes the forming, storming, norming stages of Tuckman's (1965) model, when the team comes to its performing level at the end of a project, the team often has to disband (Cornick and Mather, 1999). Project-based teams thus lack further opportunities to improve as a team and they have to go through a new 'learning curve' at the beginning of each project.

In general, teamwork studies in construction have looked into various factors that affect team performance, such as team leadership (Cheung *et al.*, 2001), team culture (Moore and Dainty, 1999), team communication (Perry and Sanderson, 1998), team evaluation (Busseri and Palmer, 2000), team members' participation (Leung *et al.*, 2004) and team design factors (Ahmad and Sein, 1997). However, the key factors that contribute to team performance are problem-solving and decision-making (Guzzo and Salas, 1995).

In construction projects, problem-solving often takes place in a team environment (Anumba *et al.*, 2001; Gunasekaran and Love, 1998). Recent construction research has focused largely on information technology (IT)-enabled collaboration systems to support team problem-solving environments in construction projects. Examples of these systems include: ADePT – Analytical Design Planning Technique (Austin *et al.*, 2002); CONSCOM – a decision-support software tool for construction scheduling and change order management (Karim and Adeli, 1999); HYCON – a virtual prototyping system to aid at the conceptual design stage (Zhang *et al.*, 2004); and CLEVER – Cross-sectoral Learning in the Virtual Enterprise (Al-Ghassani *et al.*, 2004). A detailed account of IT-enabled collaborative arrangements developed by various research projects is given in Kamara *et al.* (2000). Other hard approaches include rational models such as fuzzy-based systems (Yang *et al.*, 2001) and multi-agent systems (Anumba *et al.*, 2001) to support collaborative decision-making. However, with the complexity and messiness of construction project environments, the capability of these hard methods to capture soft issues and the applicability of these research studies in practice are major concerns (Barrett and Barrett, 2003). In fact, several studies have identified that IT-enabled collaborative environments alone are insufficient for problem-solving (see, e.g., Ingirige, 2004; Lurey and Raisinghani, 2001).

Ingirige (2004) explored the use of IT in knowledge sharing mechanisms and revealed that IT alone cannot deliver success. Lurey and Raisinghani (2001) stress that despite technology playing an important role in connecting virtual teams who are separated across space, time and/or organisational boundaries, technology alone is insufficient and face-to-face interactions are still essential to promote effective teamworking. In fact, Maznevski and Chudoba (2000) show that a rhythmic temporal pattern of interaction incidents can be developed within virtual teams by regular face-to-face meetings. Similarly, Busseri and Palmer (2000) note that team meetings are an important platform for team communication and solving project problems. Thus, understanding soft aspects of problem-solving enabled by team interactions is important before devising IT communication systems. Li and Love (1998, p. 721) confirm this when they state, 'current research in construction problem-solving has been focused heavily on developing decision aids, innovative techniques and methods for construction professionals to formulate and solve problems. There is very little research that has been done in understanding construction problem-solving as a cognitive process.'

The importance of addressing soft issues in problem situations, which arise especially through project changes, has been identified in the literature. For example, CIRIA (2001) argues that the soft issues of project change management, which cannot be captured by hard tools and methodologies, are of significant importance. Wu *et al.* (2004) explain that due to the complexity of change orders and the uncertain nature of the interaction among them, a simple statistical summation is often unable to describe fully the characteristics influencing change orders. Moore and Dainty (1999, 2001) stress that the adversarial cultures and communication issues within the construction industry affect the effectiveness of managing change and, therefore, team issues need rethinking. In their study on the project team, Sommerville and Dalziel (1998) conclude that the essential factors for effective teamworking in construction are related to participants' knowledge and the activities connected with problem-solving and team interactions.

This study attempts to understand these soft issues, in essence the team interactions and participants' knowledge flows during problem situations, specifically during reactive change situations in projects.

Unexpected changes generally become the basis of a problem within a project (Cornick and Mather, 1999). As established in the background section in Chapter 1, project changes are the most common source of disruption, disagreements, dissatisfaction and even litigation among participants in construction projects (CII, 1994). Therefore, major problem-solving activities within construction projects can be identified as 'managing change situations'. Before addressing the team interaction processes and participants' knowledge flows during change situations, the next section explores the project change management process based on previous research studies.

2.3   Construction Project Change Management

2.3.1   Change management in general

Change, as described in the *Concise Oxford Dictionary*, is the act or an instance of making or becoming different, an alteration or a modification. This is a very broad definition of change. Change at the organisational level has been extensively researched within the organisational change and organisational design literature. Organisational change can take many forms, ranging from globalisation to redesigning of firms through to changes in office administrative processes. Planned organisational change is distinguished from the change that comes about by accident or by impulse or that might be forced on an organisation (Burnes, 2000). Lewin's (1951) three-step planned change model – unfreezing, moving, freezing – provides a general framework to understanding the process of planned organisational change. However, later emergent approaches to change processes argue that change cannot be managed through such linear steps and stress the unpredictable nature of change (see, e.g., Mintzberg and Waters, 1982).

Organisational change studies (see, e.g., Weick and Quinn, 1999) describe the distinction between 'continuous, evolving and incremental' and 'episodic, discontinuous and intermittent' changes. This distinction between continuous versus episodic change exists in organisational change research in different ways – such as incremental versus radical, first-order versus second-order, and convergence versus divergence (Van de Ven and Poole, 1995; Weick and Quinn, 1999). At a project level, the 'design development' can signify the continuous change type whereas the episodic change type can occur at any stage, especially during the construction phase. Organisational inertia (Weick and Quinn, 1999), which is the inability of organisations to change as rapidly as the environment, is identified as a key concept that drives episodic change. Thus, by being responsive to the environment, construction firms can reduce the need for episodic change.

However, change at a construction project level is considerably different to these organisational change and design concepts. Although organisational change and design processes can impact and trigger project-level changes, the focus in this book is on unplanned changes that occur at the project level in the construction industry. Unplanned change can be understood through Mintzberg and Waters' (1982) concept of 'from intended to realised strategy'. An intended strategy can deviate due to emergent strategies that were not realised. Price and Newson (2003) relate these concepts to construction at an industry level. These could also be related at a project level to understand unplanned project change. Thus, a project that starts with an original contract (intended strategy) and continues its project activities based on this original contract (deliberate strategy), can deviate due to unplanned changes (unrealised and emergent strategies). Deletions and omissions from the original contract can be regarded as unrealised strategy, whereas additions to the original contract can be regarded as emergent strategy, which ultimately results in a revised contract (realised strategy).

The next section shifts from the general management literature to the construction literature to understand unplanned change within projects.

2.3.2   Nature of construction project change

CIRIA (2001, p. 10) views construction project change as 'an alteration or a modification to the pre-existing conditions, assumptions or requirements'. These project changes are additions, deletions or revisions within the scope of a project contract that can cause an adjustment to its original contract price, time and/or quality (CII, 1994). Construction literature uses contractual terms such as 'change orders' (CII, 1991) and 'variations' (Akinsola et al., 1997) to mean project scope changes that follow to implementation stage. In addition to these direct project scope changes, other elements of a project that are likely to impact on project change can be project organisation (such as team), work execution methods (such as processes and procedures), control methods (such as budget and schedule controls) and contractual risk allocation. Project change is also closely linked to other construction issues such as rework (Love et al., 1999, 2000), field rework (Fayek et al., 2003), quality deviations (Burati et al., 1992), defects (Atkinson, 1999), human errors (Wantanakorn et al., 1999), and claims and disputes (Jergeas and Hartman, 1994).

Construction change can be classified in terms of its significance, need and timing. Depending on the significance of the change it can be identified as strategic or operational (Barrett and Stanley, 1999). Further, whether or not change is optional or compulsory, it can be classified either as 'elective' or 'required' change. Based on timing, change is termed 'reactive or proactive'. This classification is parallel to the 'anticipate versus emergent change' classification. Anticipated change is planned change ahead of time that occurs as intended. Emergent change, on the other hand, is change that arises spontaneously out of local circumstances and is not originally anticipated. In the construction literature, the timing effect of change is emphasised by the 'pre-fixity versus post-fixity' classification. Pre-fixity changes are changes that occur in the design development stage, and post-fixity changes are changes that occur after the design development stage (CIRIA, 2001).

In general, the opportunity to introduce change reduces over time. Conversely, the impact of change increases when the project moves from the design to the construction phases. Lawson (1997), when explaining the design process, states that many components of the design problem do not emerge until some attempt is being made to generate solutions and, therefore, the design problem is full of uncertainties both about objectives and their priorities in the early stages. Even though later changes can be reduced significantly by proactive change management approaches, they are still common in the construction phase due to these uncertainties and incomplete project information. The later a change is introduced, the more disruption and cost will be incurred. Therefore, unavoidable changes need to be effectively managed to minimise their disruptive effects.

In addition to the design or construction phase of construction, the procurement path is another key variable that is identified in the literature. Theoretically, as CIRIA (2001) points out, the procurement route and the contract type of a project can affect the impact of change in that project. For example, a traditional procurement path with a re-measurement contract provides flexibility to make changes after the award of the contract, whereas a design-and-build approach with a lump sum contract provides less flexibility. Love (2002) predicts that the new procurement paths, which involve fast tracking and higher concurrency, are likely to create more changes than the traditional procurement paths. However, the surveys of Love (2002) and Ibbs *et al.* (2003) on the effect of procurement path on project changes reveal contradictory views. Ibbs *et al.* found that even in the case of design-and-build/lump-sum contracts, considerable changes still occur and, therefore, there is no significant effect on the impact of changes from the procurement path selection. As Ibbs *et al.* (2003) conclude, project outcomes are more dependent on the expertise and experience of the project team than the procurement route. Other variables that can influence the impact and frequency of changes are the size, type and complexity of the project (see, e.g., Akinsola *et al.*, 1997).

Following on from the above discussion, it is the project-level (operational change), unavoidable (required change), unplanned change (emergent change) occurring in the construction phase of projects (post-fixity change) that is the focus of this book. Having identified the nature of project change, the next section discusses the process of managing project change.

2.3.3   Managing project change as a problem-solving process

In understanding managing project change as a 'problem-solving' process, insights from the decision-making literature become pertinent. The following discussion takes into account three insights from this source.

First, managing change process can be explained through its input (change objectives), transformation stage and outputs (change actions) (Barrett and Stanley, 1999). The non-linear nature of this process is denoted through the feedback loop. This is essentially a problem-solving process that involves a decision to be made for change implementation (Figure 2.1).

Figure 2.1   Change problem-solving process.

According to the general management literature, problem-solving and decision-making processes overlap. However, Simon (1987) attempts to distinguish problem-solving from decision-making: the activities connected with setting goals and designing actions are usually called 'problem-solving', whereas the activities that involve evaluating and choosing are denoted as 'decision-making'. Despite this useful distinction, most literature uses both terms interchangeably. The problem-solving steps can be identified as having four principal phases (Simon, 1957): intelligence activity, design activity, choice activity and review activity. In the intelligence phase the decision environment is searched. In the design phase the decision is invented and developed and the possible courses of actions are analysed. The choice activity phase is where a particular course of action is selected. Finally, the past choices are assessed and early phases are subjected to further discussion in the review phase. Mintzberg *et al.* (1976) provide similar stages of problem-solving steps: identification, development and selection. The identification phase includes decision recognition and diagnosis; the development phase includes search and design; and the selection phase includes screening, evaluation, choice and authorisation. The common theme of these decision-making theorists (Simon, 1957; Mintzberg *et al.*, 1976) is that problem recognition should precede problem-solving. This problem identification and diagnosis stage is where organisational sense-making comes into effect (Weick, 1995). The sense-making view suggests that organisational actors first have to make sense of what is happening in their environments in order to develop a shared interpretation. This begins when there is some change or difference in the organisational environment, resulting in disturbances. As Sexton and Barrett (2003) emphasise through the concepts of 'cognitive switching gears', there must be an initial cognitive trigger that makes sense of the change situation. Thus, in managing reactive change in construction projects, the project team should identify these cognitive triggers through sense-making activities, so that the change is identified at the earliest possible time and the detrimental effects are minimised.

Second, as Mintzberg *et al.* (1976) explain, the decision-making process research falls into three main streams: individual decision-making from psychological perspectives; group decision-making from sociological perspectives; and organisational decision-making from managerial perspectives. Even though the individual and organisational decision-making research provides useful insights into the construction project change management context, given that construction problem-solving often occurs in a project team environment (see Section 2.2.4), it is the group decision-making literature that is more relevant.

Third, Li and Love (1998) explain the ill-structured nature of construction problems that involve many variables and interdependencies. According to them, the solutions to such ill-structured problems are multiple rather than a single, optimal solution. Ill-structured decisions are novel, non-programmable and unusually consequential compared to structured decisions, which are repetitive and programmable (Simon, 1957). Despite the significant amount of research on decision-making tools, for ill-structured decisions, 'intuitive' decision-making is still seen as taking precedence (Simon, 1987; Bazerman, 1994; Bennett, 1998). The ill-structured nature of problem-solving during change situations is explored in the next two sections.

2.3.4   Causes of construction project change

Project change can be triggered from internal or external causes. The external causes can emerge from wider environmental forces and the multiple-organisational level forces. In a recent study, Sun and Meng (2009) offer a similar classification of change causes. The wider environmental change causes can be from technological changes, customer expectation changes, change in competitor's activities, change in government policies, change in the economy and finally demographic changes in society (McCalman and Paton, 2000). At the multiple-organisational level, the triggers can be organisational change and design processes, such as changes in structure, production system, cultural changes and technologies (Voropajev, 1998).

Many causes of project changes are internal and generated from the intra-project environment. These can be either design-generated or construction-generated (Love et al., 1999). The examples of design-generated causes of change are design errors, omissions and change in client preferences. On the construction-generated side, examples are bad workmanship, unforeseen site conditions, poor site management and defective materials and components. A higher proportion of project change comes from the design process rather than from the construction process (Love and Li, 2000). According to Love et al. (2004, p. 427), 'design related rework in the form of change orders is the major source of rework in construction projects'. Uncertainty generated by poor information – information that is missing, unreliable, inaccurate and/or conflicting – at early stages of a project leads to the majority of reactive changes and rework. Even though the change literature identifies errors as a major cause of change, it is the managerial errors that often need more attention than the technical errors (Atkinson, 1999; Wantanakorn et al., 1999). Related to managerial errors, common change causes are poor co-ordination and integration; time and cost pressures; and inadequate skills and knowledge of participants. To this end, client role and knowledge of construction processes (Akinsola et al., 1997; Walker, 1998; Love et al., 2004) and how client needs are addressed at the briefing stage (Chinyio et al., 1998) have a significant influence over the impact and frequency of project changes.

On the whole, the origin of project-level causes such as errors, omissions and poor information can be traced back to the skills and the knowledge of project participants and how these are combined to manage construction projects (Tombesi, 2000). As Manavazhi (2004) notes, construction practitioners assess change and rework largely based on intuition sharpened by years of experience. However, the problem with the construction change management literature, as discussed before, is the tendency to prioritise the production of various tools and techniques to systemise the change process without adequately understanding the key role that the knowledge and experience of the participants play in managing projects.

2.3.5   Effects of construction project change

There are two principal types of change: change that is beneficial to a project (such as changes from value management exercises); and change that is detrimental and not value-adding to a project. In the case of beneficial change, an efficient project change management system should be in place, in order to reduce any unnecessary rework and also to feed-forward good practice to future projects. However, an effective project change management system is arguably more a priority in the case of detrimental changes, in order to minimise disruptive effects.

Changes may directly result in cost and time overruns. The major cost due to change is rework. The direct cost of rework in construction projects is considerable and has been found to be in the region of 10–15% of contract value (Love et al., 2000). In studying the magnitude of designer time on design revisions, Manavazhi and Xunzhi (2001) found that about 30% of the total planned designer time was spent on design variations, and most of these design variations were client-initiated changes.

There are also indirect consequences of change such as lower productivity. According to Bower (2000), the productivity-related indirect effect includes loss of work effort; loss of time in pausing the work; loss of float; loss of rhythm; and unbalanced work gangs. Quantitative studies show that there is a strong correlation between the amount of changes on a project and loss of productivity (Thomas and Napolitan, 1995; Hanna et al., 1999). These studies further reveal that the later the change occurs the higher will be the loss of labour efficiency.

Moreover, effects of change can take any of these forms: changed communication, changed project information, rescheduled work methods, interrupted cash flows, accelerated measures, expended additional time and cost, increased co-ordination, increased waste in abortive work, increased uncertainty and low morale (CIRIA, 2001). Ultimately, these interruptions can generate claims and disputes between the construction parties (Akinsola et al., 1997). As Jergeas and Hartman (1994) note, about 70% of contract claims can be traced to project changes in the form of design revisions, extra work and design errors. Sun and Meng (2009) generally categorise these effects that are related to contract relationships and working conditions under 'people and relationship' effects.

The next section brings together the above discussion in the form of a set of project change contextual factors.

2.3.6   Context of construction project change

Augier *et al.* (2001) show that context and contextualisation are central elements in complex problem-solving with highly constrained time frames. Thus, 'problem-solving depends both on the problem-solvers, the environment in which they exist and the emerging context in which problems become situated' (Augier *et al.*, 2001, p. 127). The context-specific factors related to project-based learning and team knowledge appear in different classifications in the literature. For example, Bresnen *et al.* (2002) group them under project characteristics, process characteristics, networking, learning capture and organisational context. Robinson *et al.* (2002) categorise them as product, process and people shaping factors. Simons and Thompson (1998) divide decision-making environments into four sets: environmental characteristics (such as economic conditions and industry conditions); organisational characteristics (such as organisation structure and culture); decision-specific characteristics (such as nature, time, risk and complexity); and individual managerial characteristics (such as values and cognitive abilities). Table 2.1 brings together these different classifications to identify a classification that suits construction project change context.

Taking these different classifications into account (see Table 2.1), the change problem-solving context is identified in this study as having four key layers: 'process characteristics', 'group characteristics', 'organisational characteristics' and 'wider environmental characteristics'.

'Process characteristics' were discussed under project-specific causes of change (see Section 2.3.4) and include design-driven triggers (e.g. design errors/omissions, uncertain information, ineffective co-ordination); also under effects of project change (see Section 2.3.5) and include task dependencies (e.g. significance, complexity, uncertainty, dependency) are grouped. 'Group characteristics' were described under the teamwork issues (see Section 2.2.4) with such factors as team leadership, team culture, team communication and team composition. 'Organisational characteristics' were explored in the organisational level causes of change (see Section 2.3.4), for example, organisation structure and organisation culture, which have complex influences over the project change management process. The 'wider environmental characteristics' represent the factors beyond the control of project participants, with potential to influence the change management process. This category includes the factors that were described under external causes of change (see Section 2.3.4) such as technological changes; customer expectation changes; change in competitor's activities; change in government policies; change in the economy; and demographic changes in the society.

The research question (RQ) and proposition (P) that generate from the above context variables are given next:

Table 2.1 Construction project change context.

Classification for project change context	Classification based on		
	Bresnen et al (2002)	Robinson et al (2002)	Simons & Thompson (1998)
Process	Process	Process *Product*	*Decision-specific*
Group ⇐	*Project Networking Learning capture*	People	*Individual-managerial*
Organisational	Organisational	*Not considered*	Organisational
Wider environmental	*Not considered*	*Not considered*	Environmental

RQ1: What are the key contextual factors of the reactive change process?
P1: The reactive change process is very complex and ill-structured rather than simple and well-structured, due to various contextual factors that could be identified under process characteristics, group characteristics, organisational characteristics and wider environmental characteristics.

Having understood the context of project change, the next section explores change management approaches.

2.3.7   Construction project change management approaches

Change management in construction is central to the project management process; indeed, most project management books allocate a complete section for changes and variations (see, e.g., Levy, 2000; Gould and Joyce, 2002; Walker, 2002). Since construction decision-making takes place in a team setting, effective project change management does not rely solely on the role of a project manager; rather, it requires appropriate input from all relevant team players. Previous approaches to construction project change management adopt a variety of different perspectives.

CII (1994) and CIRIA (2001) provide best practice guidelines on project change management. These studies suggest considering a proactive approach for detrimental change, as well as for beneficial changes. By accepting that change is completely unavoidable, the project stakeholders can reduce or mitigate the effects of unavoidable change through proactive mechanisms. The guidelines of CII (1994) and CIRIA (2001) are based on five principles:

- Anticipate change: promote a balanced change culture at the start of a project and develop a strategy, including communication and documentation policies, to deal with changes when they occur.
- Recognise change: identify and discuss potential changes before their actual occurrence.

- Evaluate change: following the recognition of change, evaluate change and identify its impact on the project in order to decide its most effective implementation.
- Resolve change: implement change, which also involves monitoring and recording the change implementation process.
- Learn from change: discuss the root causes of change and learn from mistakes for future projects.

These principles aim to mitigate the disruptive effects of change by suggesting a framework established from the start of the project to deal with change. In addition, CIRIA (2001) introduces good practices for proactive change management. For example, effective briefing as a tool to reduce required change; value management to manage elective change; risk management to forecast change; and an open project culture with free communication and continuous feedback to control the occurrence of change. Other proactive change management measures are concurrent engineering concepts (Gunasekaran and Love, 1998; Love and Li, 1998; Love *et al.*, 1998), which propose a multi-disciplinary team approach at the design phase by engaging the contractors and subcontractors as early as possible into the design solution. The LINK-IDAC research project (Cox *et al.*, 1999) proposes adopting different design management strategies, such as design for manufacture, early in the project to reduce substantial design changes that may occur at the post-contract award stage. Love *et al.* (1999, 2000) assert that by understanding causes and costs of rework, rework prevention techniques can be employed to manage project change. By reviewing a range of key publications in the construction project change management field, Sun and Meng (2009) have introduced two taxonomies for change causes and change effects, which could be used as a proactive change management guide. However, the success of the above guidelines and good practices depends on how well a project team captures and shares their knowledge and experience during implementation.

Other studies have approached change management from a process management perspective. Kagioglou *et al.* (2000), for example, introduce a separate change management process within the generic design and construction process protocol. This change process is further supported by a 'legacy archive': an IT-enabled document system to record and review project information at each phase. This approach to change management is driven by the information-processing view that believes in efficiency of processing information. In the knowledge-based era, it is the knowledge that team members bring to the pool that is more important against the general project information that could be captured by hard IT systems (see, e.g., Nonaka and Takeuchi, 1995; Leonard and Sensiper, 1998).

Drawing from this previous construction project change literature, it is evident that the change problem-solving has been viewed essentially as an information-processing activity. The evolution of knowledge-based views of the firm has lifted the understanding of problem-solving from an 'information-processing activity' to a 'knowledge-intensive activity'. This evolution of the theory of the firm is explained in the next section.

2.4   Knowledge Management Perspective

2.4.1   Evolution of knowledge-based theories

Up to the 1980s organisation theory was dominated by information-processing views, which are rooted in the works of Simon in the 1950s and followed by researchers such as Galbraith and Tushman and Nadler, in the 1970s. Simon (1957) viewed organisations as systems that process information to solve problems. Building on Simon's views, Galbraith (1974) and Tushman and Nadler (1978) viewed the organisation as essentially an information-processing entity. Thus, the information-processing perspective on organisation asserts that the key feature of organisation is to process information to enable managers to make better decisions. This recognition of information-processing needs in the organisation led to a growing emphasis on IT and decision support systems in the 1980s and early 1990s, which is commonly regarded as the 'information age'.

The assumption underpinning this perspective is that organisations should match their information-processing activities to their information needs (see, e.g., Daft and Lengal, 1986). However, empirical research has found that information processing across organisation boundaries presents significant barriers to effectiveness. Tushman (1978, p. 628), for example, emphasises that 'some project members often will not have all the appropriate information within their group, they must import information and ideas from outside the group.' This absorption of external knowledge presents a unique challenge in multi-organisational, multi-disciplined, project-based environments, as successful project delivery requires the development and application of a wide range of specialist knowledge located in different actors, and that actors mutually 'know' how their roles fit with each other. This cognitive dimension cannot be overcome by information-processing alone; rather, it is the integration of disparate, actor-specific bodies of knowledge across the inter-organisational project context that is the primary task of organisation and that determines their performance.

This limitation of the information-processing view has stimulated the development of an alternative theory of the firm, which has blossomed since the mid-1990s, that recognises that 'knowledge is the key asset' or 'knowing is the key process' (Spender and Grant, 1996; Empson, 2001) in delivering organisations' competitive advantage. This new understanding led to the argument that the focus of an organisation should be on 'enhancing organisational capabilities' rather than 'adapting to the environment' as suggested by markets/environments theories, or 'information-processing' as suggested by the decision-making theories. Accordingly, Grant (1996b) suggests that the primary role of the organisation should be the integration of knowledge: that is, by generating new combinations of existing knowledge. Kogut and Zander (1992) describe this as an organisation's combinative capabilities. Teece (1998) extends this discussion with the idea of dynamic capabilities, which is constantly reconfigure knowledge, technologies and competencies to achieve sustainable competitive advantage.

Empson (2001) identifies two perspectives of knowledge: 'knowledge as an asset' and 'knowing as a process'. From the 'knowledge as an asset' perspective, knowledge is often viewed as an objectively definable commodity, which can be managed and controlled by certain mechanisms. For 'knowing as a process' viewers, knowledge is a social construct, developed, transmitted and maintained in social situations. Dixon (2000, p. 159) provides interesting metaphors to explain these two perspectives of knowledge: 'if the warehouse is a metaphor for the stable view of knowledge, then a metaphor for dynamic view of knowledge may be water flowing across ... The warehouse image has about it a feeling of control ... the flowing water image seems less controllable, but also powerful.' Fahey and Prusak (1998) state that when knowledge is equated with information, it should not be a surprise to find it defined principally as a stock rather than as a flow. According to McDermott (1999), knowing is a human act and any discussion of knowledge is meaningless in the absence of a knower: 'the heart of knowledge is a community in discourse and sharing ideas, as such to leverage knowledge we need to focus on the community that owns it and the people who use it, not the knowledge itself' (McDermott, 1999, p. 110).

This move from asset-based views to process-based views is seen by Grant and Grant (2008) as a second generation of knowledge management. They identify a series of leading researchers in this second generation such as McElroy (2000) and Snowden (2002). McElroy (2000) draws on concepts from complexity theory to explain the complex and self-organising nature of knowledge, which cannot be simply managed by codification approaches. Supporting McElroy's (2000) arguments, Snowden (2002) shows the need to go beyond managing knowledge as a 'thing' to also managing knowledge as a 'flow' or a 'process'. Snowden (2002) asserts that knowledge can be seen paradoxically, as both a 'thing' and a 'flow' requiring diverse management approaches. To Hansen *et al.* (1999) these diverse management approaches are either codification or personalised strategies. When knowledge is seen as a 'thing', codification strategies, which especially disseminate explicit knowledge through person-to-document approaches, are considered. When knowledge is seen as a 'flow', personalised strategies, which especially disseminate tacit knowledge through person-to-person approaches, are considered.

However, as Snowden (2002) explains, rather than taking the two extremes of asset or process views, it is important to maintain an appropriate balance between the two views, depending on the cost of dissemination. Simultaneously, other researchers identify the importance of creating a balance between the core knowledge processes: 'knowledge creation' and 'knowledge transfer' (Von Krogh *et al.*, 2001). In March's (1991) terms, this means creating the appropriate balance between exploration and exploitation. The essence of exploitation is the refinement and extension of existing competencies, for which the results are proximate and predictable. The essence of exploration is experimentation with new alternatives, for which the results are distant and uncertain. Thus, organisations should balance not only between asset and process views of knowledge, but also between existing and new knowledge, in achieving sustainable advantage.

This book argues that the knowledge-based view of the firm opens new avenues to approach effective project change management in construction. Thus, the shift from thinking 'construction problem-solving' as 'information-processing' activities to 'knowledge-processing' or 'knowledge-intensive' activities is essential. The construction project management literature, in general, shows a trend towards this knowledge-based approach. The next section highlights these general knowledge-based views that appear in the construction literature.

2.4.2   Knowledge-based views in the construction literature

Looking into the next generation of knowledge management systems for construction organisations, Anumba (2009) explains the challenge in implementing knowledge management given the complexities in construction project environments. However, he shows that it is also for the same reasons that construction sector organisations cannot afford to ignore knowledge management.

Construction knowledge management literature has, in several instances, identified the importance of managing the knowledge that is generated through construction activities. For example, Winch (2002) explains that construction projects need to manage knowledge that is created throughout the construction process. Especially knowledge and learning are generated in solving problems through team discussions and dialogues during the construction process. Team learning especially takes place during regular meetings involving different participants (Huber, 1996; Barlow and Jashapara, 1998; Busseri and Palmer, 2000; Egbu et al., 2003). As Sexton and Barrett (2003, p. 615) further stress, 'innovation often takes the form of pragmatic problem-solving on site that could not have been reasonably predicted before the project started. For such problem-solving to become true innovation the solutions reached for particular problems, should be learned, codified and applied in future projects.' Therefore, 'to be successful, the project-based firms need to integrate the experiences of projects into their continuous business processes' (Gann and Salter, 1998, p. 447).

However, it is often held in the literature that the construction industry faces a knowledge diffusion problem. The knowledge generated within projects is mostly limited to individuals involved and is not widely diffused in the organisation (Winch, 2002; Barlow and Jashapara, 1998). Gieskes and Broeke (2000, p. 194) argue that 'lessons learnt may become individual tacit knowledge when they are not captured on an organisational level: there are no signs of a systematic diffusion within the organisation or across organisational boundaries.' Maqsood et al. (2006, p. 81) further describe how 'project expertise is personal and pervasively tacit. It is rarely acquired in an explicit form and hardly ever shared among others in a structured way.' This lack of learning mechanisms in generating feedback loops within projects is repeatedly emphasised in construction learning literature (see, e.g., Scott and Harris, 1998; Kululanga et al., 1999; Lu and Sexton, 2009). Price et al. (2004) argue that for construction project teams to be effective in their roles, they need to be provided with more access to a wider range of information, knowledge and appropriate technologies, which in turn demands effective learning and dissemination mechanisms. In fact, CII (1994) argues that lessons learned from the projects are often not shared effectively with project teams of future projects, and the project databases developed throughout the projects are not effectively maintained at the corporate level for future use.

Most studies identify that this ineffective learning is due to the project-based nature of construction (see, e.g., Bresnen *et al.*, 2002). Projects are problematic as they lack organisational memory and the natural knowledge transfer mechanisms that occur in permanent organisation settings. According to Hobday (2000), in construction, project-based organisation is a natural form and not a design choice. In a project-based form the project is the primary unit for production, organisation, innovation and competition, thus creating problems of inter-project and cross-project learning. This project-based arrangement therefore has pushed the construction industry away from long-term learning and has forced it to focus on short-term productivity (Miozzo and Ivory, 2000; Dubois and Gadde, 2002). For example, Disterer (2002) stresses that the project review stage often has to be dropped because of time constraints and members immediately moving on to other projects. Similar knowledge diffusion and long-term learning problems exist in other project-based settings; for example, in software engineering projects (Disterer, 2002); in engineering projects (Koskinen *et al.*, 2003); in development projects (Fong, 2003) and in financial projects (Schindler and Eppler, 2003).

The construction literature proposes different knowledge-based solutions for this knowledge diffusion problem. For example, some authors identify feedback and project review systems to manage change through past lessons (CII, 1997; Kagioglou *et al.*, 2000; Franco *et al.*, 2003). Some authors offer IT-based solutions to capture project knowledge (Kamara *et al.*, 2002; Udeaja *et al.*, 2008). Other knowledge management initiatives propose frameworks/flowcharts to measure knowledge management capability in a construction company (see, e.g., Kululanga and McCaffer, 2001; Robinson *et al.*, 2002). By introducing technology-based solutions or approaches to measure knowledge management, these authors favour an asset-view of knowledge. Bresnen *et al.* (2003, p. 159) note the privileging of the asset view by saying that 'recent work on knowledge management in the construction sector still emphasises the opportunities and possibilities opened up with the applications of such technologies [information and communication technologies].'

Bresnen *et al.* (2003) argue that there are difficulties, challenges and limitations in attempting to capture and codify project-based learning by using technological mechanisms (specifically, IT). Perry and Sanderson (1998) also note these technological limitations, and stress that design activities significantly depend on social interactions and on externalising thoughts through artefacts. They explain different types of artefacts: 'design artefacts, such as plans, models, prototypes, and visualisations, and there are procedural artefacts, which may include forms, change requests, office memos, letters, schedules, and Gantt charts' (Perry and Sanderson, 1998, p. 275). Koskinen *et al.* (2003) further emphasise these limitations by stressing that in practice a team member often relies upon other team members for knowledge and advice, rather than turning to databases and procedure manuals. According to Disterer (2002, p. 513), 'in most cases, even the place where the documentations of a specific project is stored will be unknown.' People often take advantage of databases only when colleagues direct them to a specific point. Therefore, as Koskinen *et al.* (2003) state, what is more important is knowing how to find and apply relevant knowledge efficiently. On the other hand, the application of codified knowledge in future projects is limited due to lack of details within these documents. For example, Schindler and Eppler (2003) state that the relevant project documentation is often superficial and lacks records such as reasons for failure, or how solutions were built and implemented. Disterer (2002, p. 516) has similar views in that 'documentation of projects rarely contains knowledge for following projects'. This codification process is further constrained by leadership and knowledge ownership issues that create problems as to which party is responsible for capturing and storing knowledge (Chan *et al.*, 2004).

While identifying these limitations of asset-based solutions for managing knowledge in construction projects, some authors put forward process-based views. For example, Bresnen *et al.* (2003, p. 165) reveal that 'processes of knowledge capture, transfer and learning in project settings rely very heavily upon social patterns, practices and processes in ways which emphasise the value and importance of adopting a community-based approach to managing knowledge.' Therefore, there is a greater importance in trying to develop mechanisms for knowledge diffusion that are able to replicate the social nature and dynamics of knowledge management and learning processes, as opposed to the use of technology or procedure aimed at the codification of new knowledge.

Gann and Salter (2000) maintain that tacit knowledge is extremely important within the construction environment. For example, they state that 'although many project-based activities are increasingly organised using IT systems, there is still a need for personal contact. Tacit knowledge of individuals is essential to problem-solving in projects … It involves the individuals in the life of the project' (Gann and Salter, 1998, p. 441). Therefore, learning is centred on individuals within construction projects (Dubois and Gadde, 2002). Koskinen *et al.* (2003, p. 281) capture this argument by stating that 'the fact that a great deal of the know-how required, for example, in an engineering project, is tied to knowledge that is not written in documents but realised through expertise and understanding of the project personnel, is not taken into consideration as a whole.' Thus, they suggest that it is of great importance to reinforce members' tacit knowledge through face-to-face interactions. Koskinen *et al.* (2003) view face-to face interaction as the richest medium because it allows immediate feedback so that understanding can be checked and interpretations corrected. To this end, some authors such as Grisham and Walker (2006) and Ruikar *et al.* (2009) emphasise the importance of creating and maintaining communities of practice to nurture a knowledge environment in construction. Thus, the process-based views signify that knowledge within a project and the ways of perceiving knowledge is revealed through interaction.

However, adopting and maintaining a process view of knowledge creates problems of its own, especially with respect to inter-project knowledge diffusion and learning. Some authors have identified the danger in heavily relying on individuals and on experience-based learning. For example, Gill *et al.* (2001) claim that the nature of the construction industry forces it to remain an experienced-based industry and loses opportunities for theory-based learning. Li and Love (1998) assert that construction problem-solving significantly relies on experiential knowledge, which is not codified in documents and is weakly organised in memory. According to Bresenen *et al.* (2003), the reliance on individuals and their tacit knowledge and personal skills as the mainstay of the network raises two interrelated questions about the long-term implications for project-based organisational learning. First, how is an organisation able to capture learning and deploy it over the long term, when it is often embodied in individuals and manifested in their particular expertise and range of contacts? Second, what happens when the individuals leave and take their knowledge and contacts with them? For example, Chapman (1999) identifies the highly disruptive impact when construction firms lose their key project personnel. When a team member departs in the middle of a project, the knowledge developed over the project phases will be lost. As project information is often voluminous and complex it cannot be passed totally from one individual to the next, even if there is a handing-over period. Similarly, the experience gained at a project level will be lost if/when the participants are not given an opportunity to feed it forward to the future projects (Barlow and Jashapara, 1998). Finally, Lu and Sexton (2009) reveal the 'project pull' and 'project push' knowledge spirals between the individual and the organisation, which, if effective, produce tacit, experiential knowledge accumulation and learning that provide the basis for subsequent cycles of project-based innovation.

In summary, this discussion leads to the conclusion that construction projects are very much centred on the tacit knowledge and experience of project personnel. For long-term learning, project knowledge needs to pass from the individual and project levels to the organisational level. However, the empirical research reveals the lack of learning and linkage from project to organisation level. The asset-based solutions, especially IT systems, argue for the use of codification strategies. However, as explained above, effective use and application of these systems are limited. The process-based solutions promote personalisation strategies and interactions between team members to disseminate project experience at the organisation level. As the literature suggests, the construction industry is currently biased towards this process-view of knowledge. However, process-based solutions have limitations too by placing a heavy reliance on individuals. Taking into account these different strategies and their limitations, it is important to maintain an appropriate balance between asset- and process-based solutions. Hence, the prevalent process-based practices can be enhanced by personalisation strategies; and codification strategies can be used appropriately to stimulate knowledge dissemination among individuals, so that an appropriate balance is maintained.

The knowledge-based view offers a general and a broad understanding of the role of knowledge in construction settings. However, the existing construction literature is arguably limited by not offering a deep understanding of the role of knowledge during change management situations in projects. To this end, the general knowledge management literature provides an understanding of the significant role of knowledge during problem-solving situations that are facilitated by team interactions. For example, according to Augier *et al.* (2001), when people solve problems, especially complex and unstructured problems, they bring knowledge and experience to the situation, and as they interact during the process of problem-solving they create, use and share knowledge. In Leonard and Sensiper's (1998, p. 117) words: 'in working groups, individuals from different backgrounds (cultures, experience, training, cognitive styles) draw upon their pool of tacit, as well as explicit knowledge, to contribute.' In fact, Nonaka and Takeuchi (1995) explain through the theory of knowledge conversion (discussed in Section 2.5.2) how different forms of knowledge can interact with each other and form new knowledge during shared activities. As such, this book draws extensively on the general knowledge management literature, in order to understand the role of team knowledge during reactive change process. The next section discusses these literature findings.

2.5   Role of Knowledge During Reactive Change Process

This section brings together relevant knowledge management literature to understand the significant role of knowledge during shared problem-solving, especially during the reactive change process. The bodies of knowledge brought together are shown in Figure 2.2. The figure is addressed step-by-step in the proceeding discussion, starting from the properties of knowledge and moving to knowledge creation, knowledge transfer and learning literature.

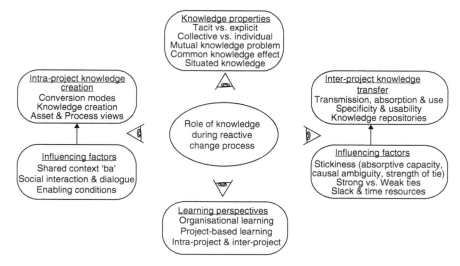

Figure 2.2   Insights from knowledge management into the reactive change process.

2.5.1   Properties of knowledge during reactive change process

Knowledge is a complex phenomenon. In Blacker's words (2002, p. 54), 'knowledge is multifaceted and complex, being both situated and abstract, physical and mental, developing and static, and verbal and encoded.' Thus, the knowledge that a project team brings to the problem situation can comprise various properties. Most literature identifies two types of knowledge: tacit and explicit. Polanyi (1966) explains tacit knowing by saying that 'we can know more than we can tell'. Tacit knowledge is highly personalised and hard to formalise, making it difficult to share with others. On the other hand, explicit knowledge is codified knowledge, which is transmittable in formal systematic language and often found in rules, policies, procedures, specifications and documents. There exist contradictory views regarding whether knowledge is both tacit and explicit; or is either tacit or explicit. To Polanyi (1966) tacit and explicit knowledge are on a continuum. Elaborating this further, Leonard and Sensiper (1998, p. 113) state that 'knowledge exists in a spectrum. At one extreme it is almost tacit, that is semiconscious and unconscious knowledge held in people's heads and bodies. At the other end of the spectrum, knowledge is almost completely explicit, or codified, structured and accessible to people other than the individuals originating it.' According to Polanyi (1966), most knowledge lies in between the two extremes, so that most knowledge has a tacit and explicit element. To Nonaka and Takeuchi (1995), tacit and explicit knowledge are alternative modes of knowing. These diverse views illustrate the multifaceted nature of knowledge.

Collins (1993) elaborates further the multifaceted nature of knowledge in four ways: encoded knowledge, embrained knowledge, embodied knowledge and encultured knowledge. Spender (1996) places these four knowledge types in a matrix by adding a second dimension to the tacit-explicit dimension. This is the individual-collective dimension. At the individual level, embrained and embodied knowledge can transpire. Embrained knowledge is contained in the physical set-up of an individual's brain that eventually shapes their cognitive abilities. This knowledge is general and transferable. Embodied knowledge is action-oriented and draws from an individual's skills that are contained in the body. This knowledge is person-specific and relevant to specific contexts. During shared problem-solving, in addition to these individual knowledge types, a collective knowledge can emerge. When people practise together they can develop common identity and mutual engagement in activities that expose them to a common class of problems that yields a store of shared collective knowledge (Sole and Edmondson, 2002). This collective knowledge can exist in two variants: encoded and encultured knowledge. Encoded knowledge is the public knowledge that could be easily transferable and is available in the wider organisation. Encultured knowledge is located in the shared practices of a social group and is generated especially through dialogue. A further separation of knowledge is embedded knowledge (Lam, 1997), which is found in organisational routines.

Collective knowledge, especially collective tacit knowledge, is emphasised in the literature as being key for knowledge creation and innovation.

For example, Spender (1996) proposes that collective (tacit) knowledge is the most secure and strategically significant. Leonard and Sensiper (1998) address this by 'tacit knowing at a group level' as opposed to Polanyi's (1966) individual tacit knowing. Leonard and Sensiper (1998) claim that collective tacit knowledge is frequently employed during shared problem-solving. Further, they explain that perhaps the purest form of collective tacit knowledge exists within teams that are characterised by complementary individual knowledge bases and bonds of shared accomplishment

Cramton (2001) addresses this collective knowledge between project team members as 'mutual knowledge'. Accordingly, mutual knowledge is knowledge that the communicating parties share in common and know they share. This can be gained through directly shared experience with other team members through socialisation activities such as visiting each other's offices and attending the same meetings. This way of gaining mutual knowledge is limited in dispersed collaborative settings. Thus, Cramton (2001) describes the 'mutual knowledge problem' in terms of communication and knowledge distribution within geographically dispersed teams. Construction project teams, which are generally geographically dispersed, therefore may face this mutual knowledge problem. Hence, communication is a recurring issue in managing knowledge within dispersed project team settings.

The group decision research addresses this 'mutual knowledge' as 'common knowledge'. This literature addresses the 'common knowledge effect' in diverse groups (see, e.g., Gigone and Hastie, 1993). Gruenfeld et al. (1996) note that during knowledge sharing between diverse problem-solving groups there is a tendency for groups to overemphasise the common knowledge that their members share and to underestimate the unique knowledge to which only one member has access. To overcome this 'common knowledge effect' and to benefit from the pooling of members' unique knowledge, Stasser et al. (1995) argue that team members need to know 'who knows what' through prior interactions. As Littlepage et al. (1997) explain, prior interactions provide not only task-experience but also group-experience. Thus, collaborative team arrangements, such as partnering, can contribute towards this. However, even though diverse groups need to know each other to utilise unique knowledge, the familiarity should not extend to a level where they no longer possess unique knowledge. The extent to which members know one another and the extent to which they hold common or unique knowledge can affect how groups process information and make decisions (Gruenfeld et al., 1996). Therefore, effective group composition strategies need to balance diversity and familiarity in order to ensure better problem-solving.

Another dimension to this collective knowledge identified in the literature, especially during shared problem-solving that takes place at site level, is 'situational knowledge'. Sole and Edmondson (2002) describe situated knowledge as embedded in work practices of a particular organisational site (locale-specific practices). Therefore, it can be seen as a special case of embedded knowledge. This occurs when individuals encounter a problem, improvise on the spot and solve the problem collectively. In construction, despite the temporary project-based nature, this situated knowledge can emerge through spontaneous problem-solving, which takes place frequently on projects.

On the whole, the concept of collective knowledge, with elements such as tacit, mutual, common, unique and situational knowledge, provides a new perspective of looking into knowledge during construction problem-solving. The research questions that generate from the above review are given below. The propositions for the research questions are developed based on the findings from knowledge-based construction literature (see Section 2.4.2).

> *RQ2: What are the properties of knowledge that the project team members use in the reactive change process?*
> *P2: The knowledge that the project team members use in the reactive change process is: more tacit rather than explicit; more collective (mutual) rather than individual; and more situational rather than prompted.*
> *RQ3: How does the project team identify and utilise this knowledge during the reactive change process?*
> *P3: Project team members, who know where the knowledge resides within the team through prior interactions, are better at identifying and actively utilising relevant team members' knowledge, during the reactive change process, compared to project team members who do not know where the knowledge resides within the team.*

The next two sections explain the knowledge creation and transfer during the reactive change process.

2.5.2   Intra-project knowledge creation via reactive change processes

As argued before, in project-based contexts, knowledge can be created through pragmatic problem-solving on site (see Section 2.4.2). When various project team members bring their tacit and explicit knowledge to a problem, this knowledge can be converted to form new knowledge through various interactions, as explained by Nonaka and Takeuchi's (1995) knowledge conversion theory.

During shared activities such as problem-solving, four modes of knowledge conversion can take place by the exchange of tacit and explicit knowledge (Nonaka and Takeuchi, 1995). They are socialisation, externalisation, combination and internalisation (see Figure 2.3). The socialisation mode emerges from tacit to tacit conversion. In this mode, tacit knowledge is

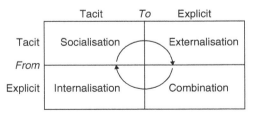

Figure 2.3 Knowledge conversion modes. Reproduced by permission of Oxford University Press, Inc.

shared directly between individuals in joint activities through observation, imitation and practice. As Nonaka *et al.* (1994) show, the socialisation mode can be evident through the presence of joint activities, apprenticeships, informal networks, social events, face-to-face settings and prior interactions. Knowledge conversion from tacit to explicit is described by the externalisation mode. This takes place in a team environment through techniques such as metaphors, analogies and models. The externalisation stage can be identified, for example, by investigating how team discussions are held. Combination mode emerges from explicit to explicit knowledge conversion. In this mode, different bodies of explicit knowledge are combined and documented through meetings, conversations and networks. Thus, combination stage can produce new manuals, documents and databases. The knowledge conversion from explicit to tacit knowledge is described by the internalisation mode. This happens when individuals re-experience others' experiences that are available in explicit forms. By identifying to what extent experimentation and reflection take place, the presence of an internalisation stage can be identified.

Nonaka (1994) explains that each of these four modes can create new knowledge independently. However, knowledge created at the socialisation and combination stages can cause problems. The 'sharability' of new tacit knowledge created at the socialisation stage is limited due to its specific nature. On the other hand, new explicit knowledge created at the combination stage will be too generic and will not extend an organisation's knowledge base. New knowledge created in the internalisation and externalisation modes, in particular, is valuable in inter-organisational contexts as it may be shared beyond the situation in which it was originally created. It is the externalisation stage that is most emphasised by Nonaka (1994, p. 19) as 'the concept of externalisation is not well developed compared to other modes. Socialisation is connected with theories of organisation culture, while combination is rooted in information processing and internalisation has associations with organisational learning.' Further to the creation of knowledge independently at each of the four modes, substantial knowledge creation happens when forming a continual cycle as shown in Figure 2.3, through the dynamic interaction between the four modes. This knowledge creation grows when it flows from individual to collective levels. As Nonaka and Takeuchi (1995) claim, this dynamic interaction of tacit and explicit knowledge, between and among individuals and groups creates a spiral effect of knowledge accumulation and growth leading to organisational knowledge creation.

However, different viewpoints exist in literature regarding the importance placed by Nonaka and Takeuchi (1995) on the externalisation stage. They necessarily take an asset view of knowledge, by emphasising that the key to knowledge creation lies in the conversion of tacit knowledge to explicit forms. Supporting Nonaka and Takeuchi's (1995) argument, Davenport and Prusak (1998) state that as difficult as it may be to codify tacit knowledge, its substantial value makes it worth the effort. Having access to tacit knowledge is insufficient; for example, when a person who possesses tacit knowledge leaves a firm, that firm will lose a part of its knowledge capital. Therefore, making tacit knowledge explicit is important. The opposing argument is that

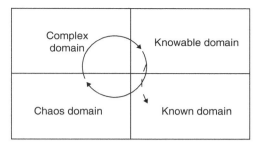

Figure 2.4   Natural flow of knowledge.

the codification of the rich (highly personalised) tacit knowledge is almost an impossible task. In the event of codifying rich tacit knowledge there will be a knowledge loss. Taking Polanyi's (1966) views, Leonard and Sensiper (1998) explain that some knowledge will always remain tacit due to reasons such as lack of motivation; lack of benefit in converting tacit to explicit; the unconsciousness of tacit knowing; and/or the difficulty in externalisation. Balancing the different viewpoints, some authors believe that a selective tacit codification process should exist, in order to share tacit knowledge across time, space and disciplines (see, e.g., McDermott, 1999).

Other process-viewers of knowledge emphasise the socialisation stage rather than the externalisation stage. According to them, the key to knowledge creation lies in focusing on informal communities of practice. For example, Snowden (2002) criticises Nonaka's knowledge conversion cycle by arguing that tacit knowledge need not necessarily have to go through a costly codification process in order to create new knowledge. Rather knowledge can be created through a natural flow in informal communities. Snowden (2002) explains the natural flow of knowledge, starting from communities that form naturally in complex domains (Figure 2.4). This complex domain with a formalisation process can form into more natural and sustainable knowable domains. A limited amount of codified knowledge can be separated from the knowable domain and be transferred to known or best practice domains. In the meantime, the knowable domain can turn back to complex domains when there may be chaos and disruption. Thus, it creates a cyclical knowledge flow that differs from that proposed by Nonaka and Takeuchi (1995).

However, studying various knowledge creation models, Grant and Grant (2008, p. 577) establish that 'the importance of Nonaka's work is evidenced by its dominance as, by far, the most referenced material in the KM field and by the number of practitioner projects implementing elements of the model. Further, while a variety of other knowledge classification systems have been proposed, variations on Nonaka's interpretation of Polanyi's original tacit/explicit knowledge concept dominate in the literature – both academic and practitioner.' However, the second-generation knowledge management researchers such as Snowden (2002) offer a more balanced view on knowledge creation by not heavily relying on the externalisation stage.

In summary, Nonaka's (1994) knowledge conversion theory offers a significant contribution to understanding of the role of knowledge during shared activities. As the knowledge-based construction literature suggests

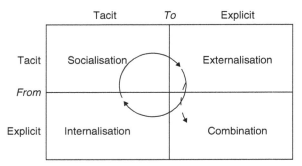

Figure 2.5   Process-based view of knowledge creation.

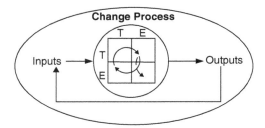

Figure 2.6   Knowledge creation around project change. T, tacit; E, explicit.

(see Section 2.4.2), construction is biased towards a process view of knowledge rather than an asset view. Thus, in order to reflect process-based views that exist in construction, this research adopts Figure 2.5, which maps Snowden's (2002) natural flow of knowledge (see Figure 2.4) on to Nonaka's (1994) knowledge conversion model (see Figure 2.3). Figure 2.5 retains Nonaka's (1994) conceptualisation of knowledge conversion modes and is built based on the assumption that these modes – socialisation, externalisation, combination and internalisation – can be approximated respectively to knowledge domains – complex, knowable, known and chaos. It therefore represents a process-based view of knowledge creation that may take place in a construction project setting, in that knowledge mainly flows between socialisation and internalisation modes (see solid line) while occasionally passing through externalisation to combination modes (see broken line).

On the whole, the change process that is represented in Figure 2.1 can be developed as depicted in Figure 2.6 in that the knowledge creation in construction that is developed in Figure 2.5 is represented as the transformation stage.

The next section describes the factors that influence this knowledge creation process.

Factors affecting intra-project knowledge creation

The knowledge management literature describes various interrelated factors that influence the knowledge creation processes. First, the presence of an enabling context, which is termed as 'ba', affects the knowledge creation process (Nonaka *et al.*, 2001). 'Ba' is a shared space in emerging relationships. However, it is not restricted to the physical 'ba'. In effect, the concept of 'ba' unifies the physical space, virtual space and the mental spaces (von Krogh *et al.*, 2000). The four most common types of 'ba' are as follows. Originating 'ba' where there are face-to-face meetings. This represents the socialisation phase and is a key to conversation and transfer of tacit knowledge. Interacting 'ba' is the place where tacit is made explicit by enabling dialogue in a team set-up. The combination of explicit knowledge takes place in a cyber 'ba'. Here, different groups work in a virtual space using collaborative environments. Exercising 'ba' represents the internalisation stage, which happens on the site, from organisation through groups to individual (Nonaka and Konno, 1998). The extent to which 'ba' is present determines the degree of knowledge conversion and creation that takes place.

Second, connected to the concept of 'ba' is the level of social interaction. Chua (2002) argues that social interaction positively influences the quality of knowledge creation. Nahapiet and Ghoshal (1998) explain that this social interaction has three dimensions: structural, cognitive and relational. The structural dimension relates to the overall pattern of connections between the actors. This includes physical and electronic resources that provide opportunities to access information and knowledge, such as workshop sessions, meetings and e-mail discussion groups. The cognitive dimension refers to resources that provide shared representations, interpretations and systems of meaning amongst parties such as shared language/codes and shared narratives (e.g. myths and stories). The relational dimension describes the personal relationships through the history of interactions between the members such as trust, norms, obligations and identity. von Krogh (1998) stresses the importance of building 'care' as part of this relational dimension. It is the relational dimension that has a greater impact on the knowledge creation process (Chua, 2002).

Third, relating to both concepts of 'ba' and 'social interaction', 'dialogue' has been identified in the knowledge management literature as the key towards creating collective knowledge through a shared context. 'Dialogue' should be open, honest, supportive and reflective. Davenport and Prusak (1998) state that in some cases there is no substitute for direct contact. Although IT is a good facilitator of data and information transmission, it cannot substitute for rich interactivity, communication and learning, which are inherent in dialogue (Fahey and Prusak, 1998). McDermott (1999, p. 104) further confirms this by arguing that 'if a group of people don't already share knowledge, don't already have plenty of contact, don't already understand what insights and information will be useful to each other, information technology is not likely to create it.'

Fourth, the presence of enabling conditions affects knowledge conversion and creation. Nonaka (1994) explains five types of enabling conditions: intention, autonomy, fluctuation, creative chaos and redundancy. Intention, autonomy and fluctuation can induce individual commitment. Intention is concerned with how individuals form their approach to the world and tries to make sense of their environment. Autonomy gives individuals freedom to absorb knowledge. Fluctuation refers to chaos or discontinuities that can generate new patterns of interaction between individuals and the environment. At an organisational level creative chaos and redundancy can trigger the process of organisational knowledge creation. Creative chaos can be generated naturally or intentionally, increases tension within the organisation, and focuses attention on forming and solving new problems. Redundancy means the existence of information more than the specific information required immediately by the individual. This may be seen as an unnecessary waste and duplication at the outset. However, redundancy of information brings about 'learning by intrusion' into the individual's sphere of perception and, therefore, is seen as crucial for the knowledge creation process.

With the influencing factors discussed above, such as 'ba', social capital, dialogue and enabling conditions, it is important to explore to what extent these factors influence the creation of new knowledge in construction project change contexts. These factors will be explored; in particular, under contextual factors relating to group characteristics (see P1 in Section 2.3.6).

Following from the insights from the knowledge creation literature, a key research question is formulated as below. The propositions and sub-propositions (SP) for the research question are developed based on the findings from knowledge-based construction literature (see Section 2.4.2).

> RQ4: How does the knowledge that the project team use in the reactive change process interact and form new knowledge?

> P4: Project teams are more likely to create new knowledge based on the existing knowledge, during the reactive change process, through a natural flow of knowledge rather than through a full cycle of the knowledge conversion process.

> SP4.1: Project teams who interact with other team members regularly through face-to-face settings, during the reactive change process, are better at utilising existing tacit knowledge and creating new tacit knowledge, compared to project teams who do not interact regularly through face-to-face settings.

> SP4.2: Project teams who actively use visualisation techniques during team discussions that arise during the reactive change process, are better at expressing and externalising their tacit knowledge, compared to project teams who do not use such techniques.

> SP4.3: Project teams are unlikely to effectively combine and codify externalised tacit knowledge arising out of team discussions in reaction change situations.

> SP4.4: Project teams are more likely to acquire superficial learning through the change experience, during the reactive change process, rather than effective internalisation through reflection.

Knowledge that is created through construction problem-solving at the project level needs to be transferred to the organisational level in order for future projects to benefit from that knowledge. This opens up the literature review to a review and synthesis of the relevant knowledge transfer literature, as presented next.

2.5.3   Inter-project knowledge transfer via reactive change process

In general, knowledge transfer refers to dissemination of knowledge from groups who have created new knowledge to the rest of the organisation. The literature refers to the terms 'knowledge transfer' and 'technology transfer' interchangeably. According to Sexton and Barrett (2004, p. 342), 'technology transfer is viewed as the movement of knowledge and technology via some channel from one individual or firm to another'. In construction project change contexts, this means dissemination to respective organisations the knowledge that is created during change events at project settings. Knowledge can be transferred from project to organisational level in this way by capturing the project knowledge that is created through problem situations and making this knowledge available to a wider community within the organisation. However, according to Davenport and Prusak (1998), effective knowledge transfer is not complete without absorption and use, following the transmission activity. Therefore, at the organisation level, not only project knowledge must be captured and made available to a wider community, but also it needs to be absorbed and re-used in future projects.

The benefits of re-using knowledge include the sharing of knowledge across a wider community, reducing duplication of mistakes and providing better understanding of problem situations (McDermott, 1999; Bhatt, 2002). However, since knowledge gains value when it is used at a specific place at a specific time (Nonaka and Konno, 1998), Sampler (1998) explains that usability of knowledge transfer depends on how specific and time relevant that knowledge is. Specific knowledge has the attributes of being possessed only by a very limited number of individuals and being expensive to transfer. Also, knowledge can be useful only during a specific period. Thus, specificity of knowledge is a key consideration in sharing knowledge in a wider community. In this context, organisations need to decide which project experience is worth transferring to a wider organisation, against the cost of transfer.

Mechanisms for this inter-project knowledge transfer are explained in various parts in the literature. Huber (1996) explains that team knowledge can be stored in three repositories for future re-use. The first repository is the minds of team members who will pass this knowledge to the co-workers of the next project. In partnering arrangements, 'serial transfer' can take place. Serial transfer involves transferring knowledge that a team has learned from doing its task in one setting to the next time that team does the same task in a different setting (Dixon, 2000). However, in the majority of cases, what happens is that a project team that gains knowledge by doing a task in one setting disbands and forms a new team in another setting. Therefore, knowledge transfer from one project to another, through project teams, is a gradual process.

The second repository is the project organisational features themselves. Grant (1996b) shows that by direction (such as directives, policies and procedures) tacit knowledge can be codified in explicit rules and instruction. On the other hand, organisational routines provide mechanisms to integrate tacit knowledge without necessarily making it explicit. In addition, the construction product itself is a repository of project knowledge.

The third repository is project files, where lessons learnt can be stored and retrieved. However, 'newly created knowledge is too complex to abstract and summarise in a file, so it is critical that lessons learned are accompanied by directory information that enables potential users to contact those who were involved in the process' (Huber, 1996, p. 72). Davenport and Prusak (1998) introduce the concept of the 'knowledge map' as a way to solve this issue, by pointing to knowledge contained within the organisation. The project review meetings followed by reports comprise a better way of preserving tacit (social) knowledge that is generated at the project level. However, these will remain local with the immediate project team. Therefore, Linde (2001) suggests that organisations need to focus on capturing this knowledge through narratives, in order for dissemination to take place to the wider community. However, Linde (2001) explains that the current trends in capturing tacit (social) knowledge (such as lessons learned systems, computer databases and video records) are not effective. The fundamental problem is in storing oral stories in a static archive. The stories will just remain local in these archives. Organisations should focus on creating occasions to share such stories and promote social interaction. These occasions can take the form of regular occasions (e.g. annual meetings, annual reviews), irregular occasions (e.g. induction, retirement) and also places (e.g. construction sites) and artefacts (e.g. albums, video). Thus, at the organisational level, construction firms need to create such occasions to transfer project knowledge to a wider audience. The use of these repositories also depends on the type of knowledge that is produced. For example, product innovation can be well recorded through design iterations and artefacts (second and third repositories), while process innovations are less likely to leave such a trail and more likely to generate tacit or informal procedural knowledge (first repository).

In summary, the knowledge transfer literature provides significant contributions in understanding the role of knowledge during shared activities. Building from these ideas, the knowledge transfer process during construction projects is represented in Figure 2.7.

The next section describes the factors that influence this transfer process.

Factors affecting inter-project knowledge transfer

Capturing and diffusing the knowledge generated on projects is highly problematic due to the idiosyncratic, complex and dynamic nature of projects. A key problem resulting from the project-based work is the transfer of learning from one project to another, or 'internal stickiness' (Szulanski, 2000), whereby each newly formed project starts anew rather than learning from what has been done previously. Hence, inter-project knowledge transfer can be affected by factors such as absorptive capacity, causal ambiguity and arduousness of the relationships. According to Cohen and Levinthal (1990), absorptive capacity is the ability of the firm to recognise, assimilate and apply new knowledge to its commercial ends. This is largely driven by the team's prior knowledge. Thus, accumulating absorptive capacity based on

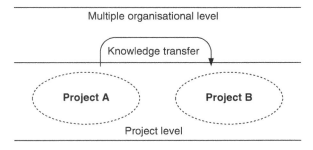

Figure 2.7   Knowledge transfer between projects. T, tacit; E, explicit.

prior knowledge will permit better sense-making (Weick, 1995) and under-standing of the situation. Causal ambiguity can be created when project team learning is unclear and complex.

Team relationships play a key role in effective transfer of team knowledge. Authors have debated this through the 'effect of strength of ties' on knowl-edge transfer and sharing (Hansen, 1999). On one side the researchers have argued that strong ties or tight couplings between people from different organisational subunits promote successful knowledge sharing, as they pro-vide redundant information (the concept of redundancy is explained in Section 2.5.2). Augier and Vendelo (1999) state that tacit knowledge is best transferred through strong ties. On the opposing side, researchers argue that weak ties or loose couplings are best for knowledge sharing, as they provide access to novel information. As Hansen (1999) explains, these contradictory views are due to a single focus on either 'search' or 'transfer' problems in isolation. A strong tie will constrain the search for novel information. This is explained by Leonard-Barton (1995) through the concept of 'core capabili-ties' that could lead to 'core rigidities'. A weak tie, on the other hand, will hamper the transfer of complex (tacit) knowledge. Thus, organisations need to balance organisational settings to address this search-transfer problem depending on the complexity of knowledge that they handle.

In bringing this understanding to construction settings, strong-coupling arrangements like partnering can constrain search and could lead to core rigidities. On the other hand, traditional procurement arrangements, which create weak ties between team members, can provide cross-fertilisation. By rotating and exposing construction participants to different members through new projects, new ideas can emerge and can provide the partici-pants with diverse rather than specialised experience. Accordingly, construc-tion firms need to exercise collaborative team arrangements cautiously to balance between specialised and diverse knowledge.

Another key factor that affects inter-project knowledge transfer is organisa-tional slack. The existence of slack resources in an organisation can particu-larly affect the absorption stages of the transfer process. Nohria and Gulati (1996, p. 1246) define slack as 'the pool of resources in an organisation that is in excess of the minimum necessary to produce a given level of organisa-tional output'. Organisational slack, as an inducement for innovation, is identified at several instances in the management literature (see, e.g., Bourgeois, 1981; Nohria and Gulati, 1996; Johnson and Scholes, 1999). As Lawson (2001) explains, although slack resources may appear inefficient in the short term, in the longer run slack will be necessary for survival and hence long-term effectiveness. Therefore, Lawson (2001) claims that it is essential to retain slack resources in an organisation, without committing all the resources of the firm for immediate output. However, it is important to understand the curvilinear (an inverse U-shaped) relationship between organisation slack and innovation (Nohria and Gulati, 1996); namely, that too much slack is bad for innovation as is too little slack. Thus, organisations need to provide slack resources for innovation in an appropriate balance.

Time that is not directly used in an organisation's primary business activity is an important source of organisational slack. Time is required to reflect and learn from experience. Lawson (2001, p. 129) emphasises this: 'just because knowledge is available, it does not mean that time is available to use it, especially in the face of cost-cutting pressures'. Thus, Lawson's (2001) argument is to allow time to think, learn and consider. In short, firms need a 'commitment of time' to think and learn despite engaging resources for their primary function. Thus, the time span between completing one project and starting a new one is a slack resource for construction firms to reflect on immediate project experience and engage in further experimentation and internalisation activities.

With the influencing factors discussed above, such as internal stickiness (absorptive capacity, causal ambiguity, team relationships), strong versus weak ties and slack/time resources, it is important to explore to what extent these factors influence the transfer of project knowledge that is created through construction project changes to the organisational level. These factors will be explored in particular under contextual factors relating to organisational characteristics (see P1 in Section 2.3.6).

Following from the insights from the knowledge transfer literature, a key research question is formulated as given below. The propositions for the research question are developed based on the findings from knowledge-based construction literature (see Section 2.4.2).

> *RQ5: How is the knowledge, which is created through the reactive change process, transferred and disseminated within the multiple organisations for potential re-use in future projects?*
> *P5: The new knowledge that is created during the reactive change process is transferred to multiple organisations, for potential re-use in future projects, through personalisation strategies rather than through codification strategies.*
> *SP5.1: The new knowledge created during the reactive change process is re-used in future projects through the individuals involved during the process, rather than through the codified documents.*
> *SP5.2: The new knowledge created during the reactive change process is disseminated and made available to the wider organisation for potential future re-use, through interactive settings between the organisational members, rather than through effective dissemination of codified documents.*

A discussion about the role of knowledge during problem-solving is incomplete without bringing in learning literature, as learning is connected to knowledge management (Pemberton and Stonehouse, 2000; Loermans, 2002; Wang and Ahmad, 2003). By identifying constructs of organisational learning as knowledge acquisition, distribution, interpretation and organisational memory, Huber (1991) confirms this link between learning and knowledge management. The next section brings in learning literature and builds up the complete knowledge-based project change process.

2.6   Towards a Knowledge-Based Reactive Change Process

Learning occurs when creating and transferring knowledge. The organisational learning research has developed in parallel to knowledge management research. The early contributors to organisational learning literature were Argyris and Schon (1978), who had developed the concepts of 'theory-in-use versus espoused theory' and 'single-loop versus double-loop learning'. In the theory-in-use, people behave consistently with their mental models whereas in espoused theory people do not act congruently with what they say (behaviour runs counter to belief system). Thus, organisational learning, according to Argyris and Schon (1978), is a change in an organisation's theory-in-use.

Argyris and Schon (1978) further argued that learning occurs under two conditions. First, learning occurs when an organisation achieves what it intended; that is, there is a match between intentions and outcome. Second, learning occurs when a mismatch between intentions and outcome is detected and corrected. Depending on the surface and deep levels of learning, two learning strategies exist, as single-loop (adaptive learning) and double-loop learning (generative learning). Single-loop learning occurs when an error is detected and corrected without questioning or altering the underlying value systems. Double-loop learning occurs when mismatches are corrected by first examining and altering the underlying value system, and then the actions. For innovation to occur it is the double-loop learning rather than the single-loop learning that should take place. According to Wang and Ahmad (2003), organisations should further consider a higher level or third-order learning called triple-loop learning, which focuses on organisations having the capacity and capability to learn continuously.

Moving beyond the above discussed adaptation theories, to include cognitive change as well, Fiol and Lyles (1985) view organisational learning as the process of improving actions through better knowledge and understanding. This includes alignment with the environment through learning, unlearning and relearning based on past behaviour. Organisational learning is more complex and dynamic than a mere magnification of individual learning (Kim, 1993). Knowledge is neither completely stored into individuals nor into the organisation (Bhatt, 2002). A part of knowledge is stored in individuals and a part of it is stored in the organisation. Individual

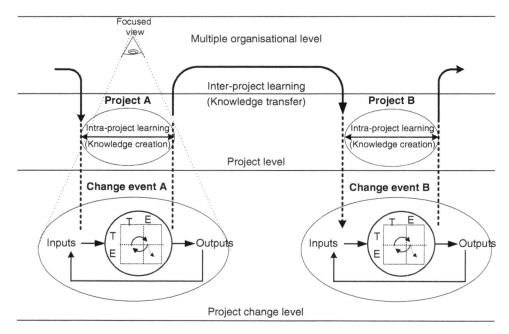

Figure 2.8   Knowledge flows between change events.

learning is an area that had developed in parallel with organisational learning. For example, Kolb's (1984) learning cycle explains how individuals
learn from their experience. Kolb (1984) stresses that reflection after experience is paramount in order to learn from past lessons. Schon (1983)
describes how practitioners could reflect based on their tacit knowing.
These views on experiential learning are similar to Nonaka's (1994) internalisation mode.

'Intra-project learning' and 'inter-project learning' are two concepts that
relate to project-based organisations (see, e.g., Gieskes and Broeke, 2000;
Robinson et al., 2002). 'Intra-project learning' is the acquisition and use of
knowledge within the project whereas 'inter-project learning' refers to the
transfer of knowledge from one project to another within the same or different time frames. These concepts can be related to 'intra-project knowledge creation' and 'inter-project knowledge transfer' respectively. Figure 2.8
builds up the broader view of knowledge flows in construction settings by
integrating the understandings gained from knowledge creation literature
(see Section 2.5.2); knowledge transfer literature (see Section 2.5.3); and the
project-based learning literature discussed above. Accordingly, Figure 2.8 is
produced by combining Figure 2.6 on intra-project knowledge creation and
Figure 2.7 on inter-project knowledge transfer.

The conceptual model given in Figure 2.9 is developed by considering a
focused view of a change event (e.g. see the focused view of Change event A
depicted in Figure 2.8) to represent the knowledge-based reactive change
process; namely, management of unplanned change in the construction

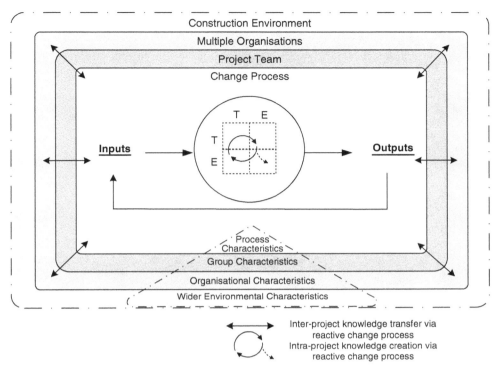

Figure 2.9   Knowledge flows during reactive change process: conceptual model. T, tacit; E, explicit.

phase within collaborative team settings. In brief, Figure 2.8 provides the broader view of the research problem, while Figure 2.9 provides the focused view of the research problem.

The core of the model represents the change process as an input-transformation-output model. Knowledge creation around a change event is represented in the transformation stage of this change process. The project-to-project knowledge transfer is represented by arrows that link a project to the multiple organisation layer through the project team layer. The context of this problem statement is shown in four layers: change process; project team; multiple organisations; and construction environment. The characteristics corresponding to each layer are shown. The change process in the inner layer is influenced by the process characteristics. The second layer represents the construction project team, which is influenced by group characteristics. The third layer represents multiple organisations, which are influenced by organisational characteristics. The outer layer shows the construction environment, which is affected by the wider environmental characteristics. These contextual factors in terms of the four characteristics are depicted in the model by a triangle to represent the direction of the impact. In summary, the model depicts the role of team knowledge during change process in construction team settings.

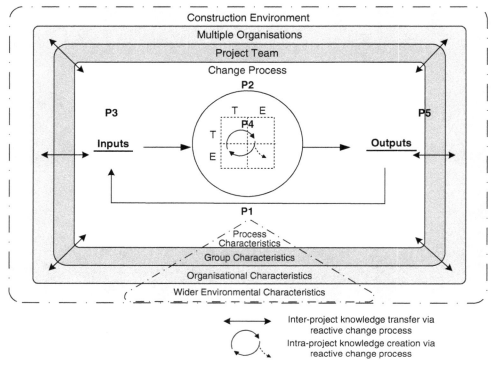

Figure 2.10   Mapping propositions (P1 to P5) within the conceptual model. T, tacit; E, explicit.

The main propositions are mapped onto the conceptual model represented in Figure 2.9 (see Figure 2.10). As indicated in Figure 2.10, proposition P1, which describes contextual factors, is represented in the triangle of characteristics. Proposition P2, which deals with knowledge properties, is mapped onto the model at the transformation stage. Proposition P3, which relates to knowledge identification and utilisation, is represented at the inputs. Proposition P4, which explains the intra-project knowledge creation, is mapped within the transformation stage of the change process. Finally, proposition P5, which deals with inter-project knowledge transfer, is mapped after the outputs.

2.7   Summary and Link

This chapter has presented key issues from the literature. Previous approaches to managing project change, or managing knowledge in general within construction project settings, adopt information-processing or asset-based views of knowledge. Thus, the construction literature is arguably limited in appreciating the significant role of knowledge as a social process during construction project settings, in particular in managing change. This led to the exploration of the general knowledge management literature, which offers a deep understanding of the significant role of knowledge during problem situations that is facilitated by team interactions. Based on this knowledge-based perspective of the reactive change process, research questions and propositions for this book were articulated. The next chapter describes the research methodology used to investigate these research questions and propositions.

Research Methodology

3.1 ## 3.1 Introduction

Chapter 2 discussed the key research issues through a review and synthesis of the relevant literature. The aim of this chapter is to describe the research methodology employed in this study. The chapter is structured as follows:

- First, the case study research design adopted in this study is described (see Section 3.2).
- Second, the data collection process is briefly explained (see Section 3.3).
- Third, the data analysis process is outlined (see Section 3.4).

3.2 Case Study Research Design

The case study methodology was used. According to Yin (1994, p. 13), the case study method is 'an empirical inquiry that investigates a contemporary phenomenon with its real-life context, especially when the boundaries between phenomenon and context are not clearly evident … the method will rely on multiple sources of evidence and development of prior theoretical propositions.' Yin (1994, p. 1) further states that 'case studies are the preferred strategy when "how" or "why" questions are being posed, when the investigator has little control over events and when the focus is on a contemporary phenomenon within a real-life context.'

The unit of analysis and criteria for case selection were considered as follows.

3.2.1 Unit of analysis

Miles and Huberman (1994) define the 'unit of analysis' or the 'case' as a social phenomenon occurring in a bounded context. The purpose of defining the unit of analysis is to identify the 'focus' or the 'heart' of the study

Managing Change in Construction Projects: A Knowledge-Based Approach, First Edition.
Sepani Senaratne and Martin Sexton.
© 2011 by Sepani Senaratne and Martin Sexton. Published 2011 by Blackwell Publishing Ltd.

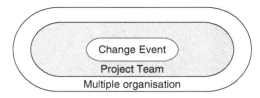

Figure 3.1   Unit of analysis.

with its boundary. The main unit of analysis in this study was the reactive change event during the construction phase within the context of collaborative project teams. The unit of analysis was then built outwards (Figure 3.1). The stakeholders (project team) around the fixed issue who are from multiple organisations were considered as the higher levels of the analysis.

3.2.2   Case screening and selection

The population selected in this research comprised the unplanned change events in the construction phase within collaborative team settings that include partnering and design-and-build (D&B) approaches. The sampling of cases from the chosen population is unusual when building theory from case studies. Random selection is neither necessary nor even preferable in the case study method as opposed to quantitative approaches, which call for statistical sampling. Thus, cases should be selected on the basis of theoretical sampling (Yin, 1994), which focuses on theoretically significant cases such as representative, disconfirming and/or extreme instances. Thus, adopting this theoretical sampling, representative cases were selected for in-depth investigation.

Based on the above selected population and theoretical sampling, the following sampling strategy was used to identify the cases. In selecting the 'multiple organisation sample', the degree to which the case study firms engaged in collaborative arrangements was the defining criterion for selection. Collaborative arrangements used in this research were gleaned from the literature, and include partnering and design-and-build arrangements. In selecting the 'project team sample', active project team members around a change event were considered. Change events were selected solely on the basis of project participants' perception of the degree of impact on the construction phase (be it time, quality or cost impacts).

Multiple cases were selected to provide multiple sources of evidence and potential replication of findings. According to Yin (1994), multiple cases should be carefully selected to predict either similar results (literal replication) or contrasting results (theoretical replication). Further, in deciding the number of cases, the richness of data, time and resource availability need to be taken into account. This study selected three 'change events' within two 'project organisations' with the aim of predicting similar results (see Section 3.3.3 for details of the selected sample).

3.3   Data Collection Process

The next step of the research method was to decide on research techniques for data collection.

3.3.1   Selection of data collection techniques

Denzin and Lincoln (2000) compare and contrast the effectiveness of methods available for data collection: observation techniques raise questions of observer bias and setting bias; documents surveys are limited as a data collection method because documents cannot respond directly to research questions; and interviewing is viewed as a powerful data collection method (Fontana and Frey, 2000). Interview techniques suit case study research as they deal with real-life settings. According to Yin (1994), human affairs should be reported and interpreted through the eyes of specific interviewees, and well-informed respondents can provide important insights into a situation. Interviews can be structured, semi-structured or unstructured depending on whether the interviewer is neutral or actively involved in the process. Denzin and Lincoln (2000) argue that interview (semi-structured and unstructured) is not a completely neutral tool, and therefore it can be influenced by the interview situation and the personal characteristics of the interviewer and interviewee (Alvesson, 2003). Thus, it is important to address these weaknesses of different methods when selecting a data collection technique.

Triangulation of data collection methods was used to mitigate against the weaknesses of using individual collection methods in isolation (Yin, 1994). Triangulation provides stronger substantiation of constructs and hypotheses, by overcoming weaknesses of each method when used alone (Jick, 1979). Despite the possibility of discrepancies that can result in the use of multi-method results (Easterby-Smith *et al.*, 1991), Jick (1979) argues that divergence can often turn out to be an opportunity for enriching an explanation.

In terms of source triangulation, while semi-structured interviewing was selected as the main data collection technique, other techniques such as document survey (e.g. survey of minutes of meetings and change records) and observation techniques (e.g. participating in project team meetings and informal observations through visiting interviewees' offices and sites for interviews) were used to some extent. In terms of time triangulation, this study undertook data collection over a period rather than a single moment (e.g. two-stage interviewing, periodic workshops involving interviewees), so that true longitudinal data could be collected where possible. In terms of methods triangulation, two data analysis approaches were triangulated, as explained in Section 3.4. Having selected interviewing as the main data collection technique, the next section describes the design of interview guidelines and the interview process used in this study.

Table 3.1   Structure of interview guidelines.

Interview section	Literature issues addressed	Relevant sections of Chapter 2
1.0 Background	▪ Nature, causes, effects, role of participants	2.3.1, 2.3.2, 2.3.3, 2.3.4, 2.3.5
2.0 Practices	▪ Communication methods, resources	2.3.7
3.0 Variables	▪ Process characteristics: significance, complexity, uncertainty	2.3.6, 2.3.4 2.2.2, 2.2.3
	▪ Group characteristics: language, distance, consistency, experience, working relationships, level of care, norms	2.2.4, 2.5.2
	▪ Organisational characteristics: organisation structure, culture, top management support, enabling conditions	2.2.1, 2.3.4, 2.5.3
4.0 Role of team knowledge during change	▪ Knowledge capture: properties, prior preparation, identification	2.5.1
	▪ Socialisation: socialisation activities	2.5.2
	▪ Externalisation: using visualisation techniques	2.5.2
	▪ Combination: refer documents, codification	2.5.2
	▪ Internalisation: team learning	2.5.2
	▪ Knowledge re-use: use in other projects	2.5.3

3.3.2   Interview structure

Interview guidelines should be structured around the argument and should not generate information overload. Thus, the interview guidelines of the research study that is reported in this book, as explained in this section, were designed to capture data around the research problem. Table 3.1 explains the development of interview guidelines with reference to the literature issues discussed in Chapter 2. The interview guidelines (see Appendix A) comprised nearly 60 questions that were structured in four sections. The first section addresses the background information of the selected change event. The second section addresses the project change management process. The third section explores the variables that affect this change event, in terms of process, group and organisational characteristics. The wider environmental characteristics were intended to be captured through section one under causes of change. The fourth section of the interview guidelines covered the role of knowledge during the change event. This section was divided into six areas: knowledge capture, socialisation, externalisation, combination, internalisation and knowledge re-use. The first three sections basically addressed proposition P1, while the fourth section addressed propositions P2 to P5.

The interview guidelines were revised following the pilot interviews to address inconsistencies that were identified, through the initial analysis of data. As Eisenhardt (1989) states, a striking feature of theory building from case study research is this frequent overlap of data analysis with data

Project A – Supermarket store refurbishment and extension

The first case study project selected was on partnering arrangement. From the short-listed change events the floor change event was selected.

Case A1 – Floor change

The interview sample included the D&B Contractor, Architect, Client Agent and the Client. Other data collection methods included participation in design team meetings and document survey of change records, minutes of meetings and documents available in the project information channel.

Project B – School project

The second case study project selected was on a Design and Build arrangement. From the short-listed change events the whiteboard change and radiant panel change events were selected.

Case B1 – Whiteboard case

The interview sample included mainly the D&B Contractor and Architect. Only partial information was received from the M&E Engineer. Other data collection methods included participation in design team meetings and document survey of change records, flow charts and minutes of meetings.

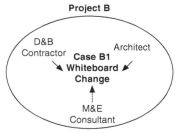

Case B2 – Radiant panel

The interview sample included mainly the D&B Contractor and M&E Engineer. Only partial information was received from the Architect. Other data collection methods included participation in design team meetings and document survey of change records, flow charts and minutes of meetings.

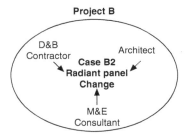

Figure 3.2   Interview sample.

collection – in Glaser and Strauss's (1967) terms 'constant comparative analysis' – and the flexibility to adjust the data collection techniques. However, the overall structure and questions were not significantly revised in order to maintain consistency across cases.

3.3.3   Interview process

First, the general project information and change events were gathered through initial interviews with one of the key participants. Based on this information a specific change event was selected and the interview guidelines were e-mailed to the interviewees prior to the interview. Subsequently, interviews, 2 to 3 hours in duration, were conducted with each active participant. The interview sample is given in Figure 3.2.

The data collected through the interviews were tape-recorded and also complemented by note-taking. The interviews were transcribed (see Appendix B for an example interview transcript).

To address ethical issues during the interview process, permission to tape-record was acquired from the interviewees before the interview. To maintain confidentiality the actual names of the projects and the participants' names were not disclosed. The transcriptions were sent back to the interviewees for their acceptance and confirmation. The next section explains the data analysis process of this research.

3.4   Data Analysis Process

First, the primary data were subject to code-based content analysis to identify key concepts and issues. Second, the relationships between the concepts were identified by use of a popular mapping technique called the cognitive mapping technique. Therefore, the content analysis in this research was used for data reduction. According to Miles and Huberman (1994), data reduction is the process of selecting, focusing, simplifying, abstracting and transforming data that appear in written-up field notes or transcriptions. On the other hand, cognitive mapping in this research was used for data display. Miles and Huberman (1994) explain data display as an organised, compressed assembly of information that permits conclusion drawing and/or action taking. The analysis was aided by computer software- NVivo for content analysis and Decision Explorer for cognitive mapping.

The above data analysis process was followed for each case study, and the relevant coding structures and cognitive maps are presented along with the research findings in Chapter 4.

3.5   Summary and Link

This chapter has presented and justified the methodological approach used in this study. The next chapter presents and analyses the research findings from the case studies.

4

Case Study Results

The aim of this chapter is to present the research findings of the individual case studies. The chapter is structured as follows:

- First, Project A with a main change event is presented and analysed (see Section 4.2).
- Second, Project B, which combines two mini change events, is presented and analysed (see Section 4.3).

4.2.1   Case study description

The case study project comprised a supermarket store extension and refurbishment. The project used a design-and-build (D&B) procurement path and was one of a series of projects that the client and the project team contracted on a partnering arrangement. The project duration was 29 weeks, and the contract value was £7 million. The selected change event was a 'change in the store flooring design'.

The original floor design of the store required a shutdown of the store. The client, however, realised belatedly that it had not considered the loss of six weeks' trading. The client, therefore, wanted the D&B contractor to change the original design and consider other floor design options that would minimise the closure of the store. The generation and evaluation of floor options took four months at the initial stage of the construction work. The team finally decided to tile the floor during night shifts and only where the floor was in bad condition, instead of the original option of replacing the terrazzo floor, which would have required the complete closure of the store.

Managing Change in Construction Projects: A Knowledge-Based Approach, First Edition.
Sepani Senaratne and Martin Sexton.
© 2011 by Sepani Senaratne and Martin Sexton. Published 2011 by Blackwell Publishing Ltd.

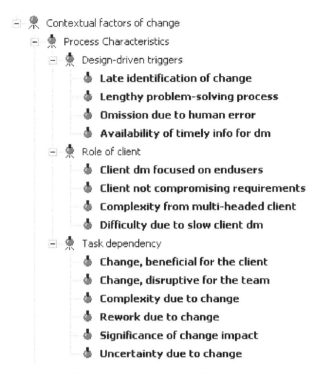

Figure 4.1 Project A: coding structure of process characteristics. dm, decision-making. Screen shot courtesy of QS International.

4.2.2   Contextual factors of change

Contextual factors of change are discussed in this section under process characteristics, group characteristics and organisational characteristics. Finally, a summary of the findings relating to contextual factors and comments on proposition P1 is given.

Process characteristics

In this section, the coding structure (Figure 4.1) and the cognitive map (Figure 4.2) that relate to process characteristics are presented. These form the basis for the following discussion, which is described under three headings: design-driven triggers, role of client and task dependency.

Design-driven triggers
The cause of this change was an omission made by the client party. The client had not experienced a terrazzo floor replacement before and was not used to refurbishing floors during occupancy. The client wanted a terrazzo floor replacement for this store, which would have needed a six-week store closure. The team made the client aware of this at the initial design stage and the client

initially agreed to the store closure. The client, however, subsequently realised that it had not considered loss of trade for this period, which it could not afford. By this stage the construction work had already started. The uncertainty of this change in client requirements was further amplified by the search for an alternative design solution taking an additional four months.

Role of client

The role of the client was identified as a key variable in managing this change. In this change event, the client's failure to notice the business aspect of the store closure in the first place caused the major change. Further, the client's delay in making a decision on the change options adversely affected project progression. The client representative himself admitted this when he stated that:

> we were very slow in deciding, as always and we do so because we can't … Yes, we took our time about it.

A key difficulty was making the client understand that it could not achieve a high-quality floor without time and cost implications.

As the architect stressed, 'the fact that they are a multi-headed client makes the client decision-making process complex'. The change could therefore be initiated by several members in the client team; for example, it could be by the store planner, the store manager or the regional team. From the client team perspective it was the store manager who mediated between the project team and the client team. However, he was given insufficient authority to make decisions and, as he perceived it, the internal decision-making process was too political and hierarchical. The store manager was frustrated in trying to get quick decisions from higher-up in the multi-headed client hierarchy.

It was apparent that the client's key focus throughout this change event was to cause minimum disruption to customers and, in so doing, reduce the loss of trading. In the store manager's words: 'because this is a trade store we wanted to solve this problem with minimum disruption to our customers'. Further evidence that showed care for customers was taking customers' complaints into consideration during this change process.

Task dependency

This change affected parties at different scales. To the client, the change was beneficial as, at the end, it received a quality floor, on time, at a reasonable price. In fact, by not selecting the terrazzo option, it made an overall saving. As the client agent explained, 'by the end we made a happy client with a satisfactory solution'.

The change was, however, disruptive from other project members' viewpoints. The change involved more co-ordination and management. The D&B contractor said the final solution was very disruptive as it had to work on a live project that involved night work. The D&B contractor further explained the disruption:

> For subcontractors, the change was disruptive more in terms of uncertainty. This uncertainty affected them in organising their labour.

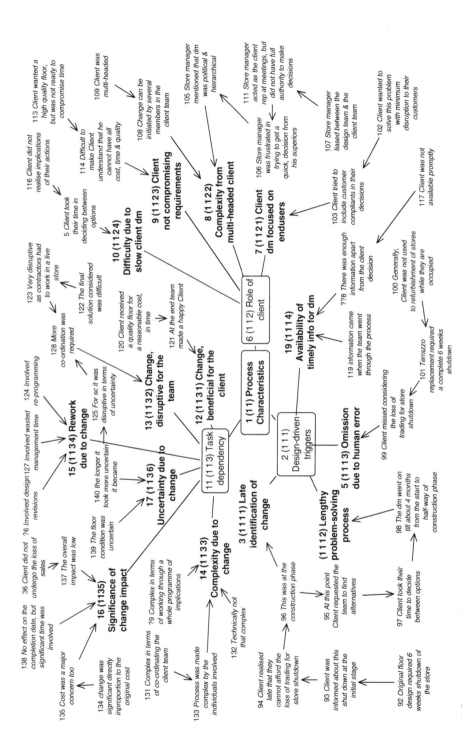

Figure 4.2 Project A: cognitive map of process characteristics.

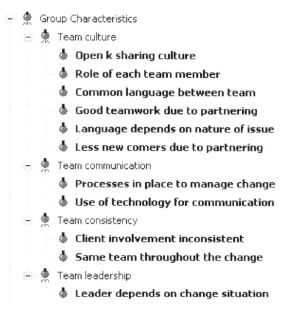

Figure 4.3 Project A: coding structure of group characteristics. k, knowledge. Screen shot courtesy of QS International.

The client agent said because the issue went on without a decision for a long time, the more it approached the end date the more uncertain they felt. The change created further uncertainty due to the unknown floor condition. Hence, this change created rework in terms of design revisions, wasted management time and re-programming.

When the team was queried about how complex they found this change event, team members said that it was not so much 'technically' complex, but 'socially' complex due to the individuals involved, especially within the client team. Further, the D&B contractor stated that it had found it complex, especially in terms of working through a whole programme of implications. The team also expressed the significance of this change in terms of time and cost.

In terms of information availability, the D&B contractor stated:

> there was not sufficient information available to make this decision initially. That is why we had to go through the process. Information came when we went through the process.

Group characteristics

The coding structure (Figure 4.3) and the cognitive map (Figure 4.4) that relate to group characteristics are presented in this section. These form the basis for the following discussion, which is described under four headings: team culture, team communication, team consistency and team leadership.

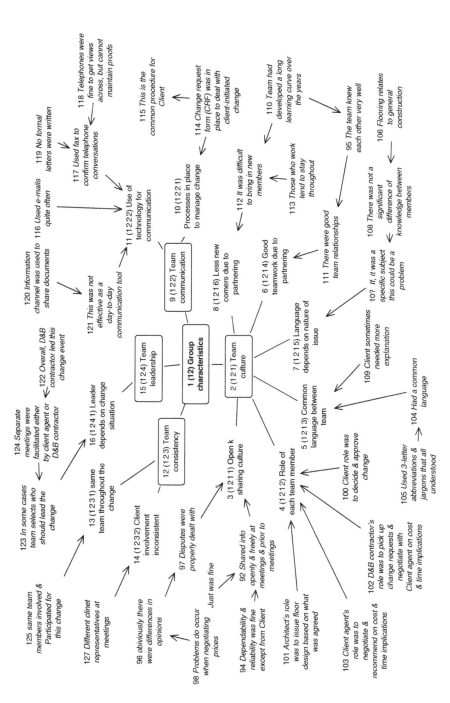

Figure 4.4 Project A: cognitive map of group characteristics.

Team culture

Project team members had different roles in managing this change event, but their roles were interdependent on each other. It was evident that the collaborative team approach, in this case the D&B path and especially the partnering arrangement, had created good teamworking and a long learning curve between the parties. The prior relationships enabled them to share information openly and freely prior to and during meetings. When the team members were queried about the trust, dependability and reliability during the process, they agreed that these existed between the parties at a satisfactory level. When there were differences in opinions they were brought to the forum and properly dealt with. The long-term team relationships further created a common language between the parties. According to the D&B contractor, they were very much used to three-letter abbreviations and jargon. Nevertheless, the architect pointed out that language could still be a problem in the case of very specific technical issues, such as refrigeration. The D&B contractor also identified a problem caused by the partnering arrangement:

> Due to the long learning curve that the team had developed over the years, it is difficult to bring in new members to the projects and those who worked throughout tend to stay.

Team communication

There was a formal change management procedure in place, called the Change Request Form (CRF), to deal with client-initiated changes. First, the change was notified by the client to the D&B contractor, who picked up the change request, evaluated it in terms of time and cost, and forwarded it to the client agent. Second, based on the client agent's recommendations, the client approved the change. During this process the team used common technologies such as e-mails, telephones and faxes to communicate. In addition, there was an information channel to share project documents. Nevertheless, as the D&B contractor explained, 'it is not so much a straightforward process as it sounds. It involves several cycles of negotiations.' These negotiations and change evaluations mostly took place at team progress meetings.

Team consistency

The team membership was consistent throughout the change event. With the exception of the client, the same individuals from the multiple organisations participated in the discussions. As the client agent described, 'the only inconsistency was the changes in client representatives, which delayed their decision'.

Team leadership

In terms of leadership, the D&B contractor stated:

> in some cases we select who should lead the change. In this, we led the change as it involved technical issues.

Figure 4.5   Project A: coding structure of organisational characteristics. KC, Knowledge Creation. Screen shot courtesy of QS International.

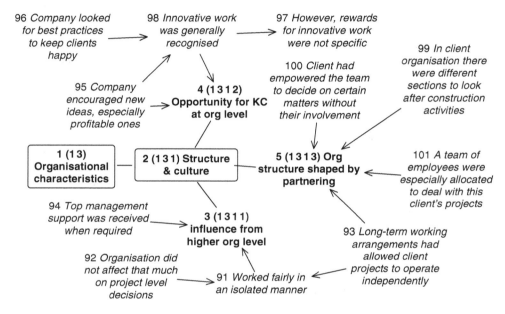

Figure 4.6   Project A: cognitive map of organisational characteristics. KC, Knowledge Creation; org, organisation.

Thus, the chosen leader was contingent on the type and scale of change.

Organisational characteristics

The coding structure (Figure 4.5) and the cognitive map (Figure 4.6) that relate to organisational characteristics are presented in this section. These form the basis for the following discussion.

Structure and culture
It was evident that the project organisation structure had been shaped by the long-term partnering arrangement. The D&B contractor, for example, stated:

> the long term working arrangements that had been developed over the years had allowed the client projects to operate independently.

and the architect confirmed this when he stated:

> We have refined the organisation structure to deal with client's projects over the years. There are two project architects who are always assigned for supermarket projects.

The client agent, too, agreed with this practice: 'there is a team of people at our firm who generally works in client projects'. In the client organisation there were different sections to handle various construction activities. For example, it had a procurement section to look after equipment; a property management section to look after the contractors and the projects, and a financial section.

Project-level decisions took place generally within the project team, with little input from senior management within the participating companies. The top management got involved only when requested by the team members. The D&B contractor stated, for instance:

> We work in a fairly isolated manner. I directly report to the board of directors and I am empowered to make decisions then and there.

The architect further explained this: 'top management involves only in case of a faulty situation'.

The team members were queried on the opportunity to create new knowledge at the multiple-organisation level. They said that their companies encourage employees to come up with new ideas and disseminate good practices. However, rewards for such practices were not explicit. The client agent described this:

> Our company is keen on new ideas if they are a new way of making money. They look for best practices to keep the client happy. Rewards are not specific. But with a new idea if you generate money and work for the company you are recognised and rewarded.

Comment on proposition P1

> P1: *The reactive change process is very complex and ill-structured rather than simple and well-structured, due to various contextual factors that could be identified under process characteristics, group characteristics, organisational characteristics and wider-environmental characteristics.*

The key process characteristic that affected this change situation was a requirement omission by the client at the design stage. In terms of task dependency, the uncertainty created by the change disrupted the project team and incurred rework. It was the 'social process' complexity rather than the technical complexity that affected this change. These findings show the potentially significant impact on project change of design-driven triggers and task dependencies that could be grouped under process characteristics (see Table 4.1).

Table 4.1   Project A: evidence relating to proposition P1.

Attributes	Evidence	Confirmation/falsification
Process characteristics	▪ Design-driven triggers, such as omissions ▪ Client-initiated change ▪ Task dependencies, such as uncertainty and social process complexity	
Group characteristics	▪ Team culture with knowledge-sharing and good team-working ▪ Team communication with change management procedures, regular meetings and IT ▪ Team inconsistency with different client representatives ▪ Team leadership that depended on the issue	Confirmed
Organisational characteristics	▪ Flexible organisation structure ▪ Learning and knowledge-sharing culture	

The case study findings revealed that the partnering arrangement appeared to have influenced a team culture that was characterised by good teamworking and an open knowledge-sharing environment. Team communication was enabled by well-established change management procedures (e.g. CRF system), regular meetings and IT. Team inconsistency in terms of different client representatives at project meetings had affected the change decision-making process. Further, the nature of change had affected team leadership during a change event. These findings show the significant impact on project change from variables connected to group characteristics (Table 4.1).

The data further suggested that the organisation structure and culture were made flexible by the partnering arrangements to deal with project-level changes. The learning and knowledge-creating culture at the organisation level was not motivated by explicit rewards (Table 4.1).

Taking these findings as a whole, it is difficult to establish clear relationships between process, group and organisational characteristics, as these factors had a complex influence on project change. Some factors enabled effective management of change while others had a negative impact. Therefore, these findings confirm proposition P1, namely that the reactive change process is very complex and ill-structured rather than simple and well-structured, due to various contextual factors that could be identified under process characteristics, group characteristics and organisational characteristics. The case study did not provide clear evidence for wider-environmental characteristics.

Figure 4.7 Project A: coding structure of knowledge properties. K, Knowledge. Screen shot courtesy of QS International.

4.2.3   Knowledge properties during change

Figure 4.7 presents the coding structure in relation to 'knowledge properties'. This is discussed in three parts below. Finally, an overall summary and the comment on proposition P2 are provided.

Tacit versus explicit knowledge

The cognitive map (Figure 4.8) that relates to 'tacit versus explicit knowledge' is presented in this section, along with the following discussion on the case study findings.

One of the key observations of the case study was that the team members relied heavily on their previous experience (tacit knowledge), rather than referring to codified documents (explicit knowledge), to solve problems connected with the reactive change.

The team were experienced in construction work in general and more specifically in supermarket projects. The fact that the multiple organisations were in a partnering arrangement allowed the same team members to work in similar types of projects with the same client, thus enabling them to build up this specific experience. The D&B contractor stated:

> if we are in the construction sector, we all have technical experience. We normally promote to the management hierarchy through this technical experience.

This long-term experience facilitated team members to successfully share their tacit knowledge between each other. The architect confirmed this: 'quite frequently, we shared lessons learnt from previous work experiences'. The extent of using past experience directly can be made difficult due to new

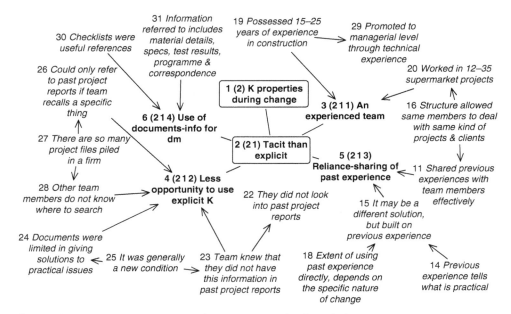

Figure 4.8   Project A: cognitive map of tacit versus explicit knowledge.

conditions and the specific nature of change. The team, however, built upon its past knowledge to come up with a new solution. This was evident in the D&B contractor's statement:

> What we are trying to do is to use the information that is available. The history and our experience tell us what is achievable. You can come up with a new idea, which is not practical. So you always have to relate old knowledge with the new condition, to come up with a practical solution. We are dealing with an existing floor. So we did not collect various documents and develop the idea, rather used our experience and the existing knowledge to the practical situation.

Therefore, the team had limited opportunity to refer to codified documents that contain past project reports. In effect, team members did not look into past project reports as they knew that they did not have a 'ready-made' solution that could be applied directly to this new situation. The use of codified knowledge was limited as construction often deals with practical issues. Further, as the client agent pointed out, 'there are so many project files in the firm and other members simply will not know where to look for', and they even 'go back and refer [to] project files, if we [they] recall a specific thing that relates'. Despite these limitations, to some extent the team referred to information that was contained in documents, such as checklists, material details, specifications, test results, programmes and other correspondence.

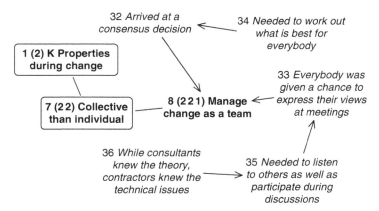

Figure 4.9   Project A: cognitive map of collective versus individual knowledge.

Collective versus individual knowledge

The cognitive map shown in Figure 4.9 forms the basis for the following discussion.

It was evident that the change event was managed as a team and that the team members were dependent on each other in managing the event. As the architect expressed it:

> we know the theory and they know the technical matters. So it is very important to relate and listen to other parties.

At meetings, when the change issue was discussed, everybody was given a chance to express their views. The D&B contractor said at these meetings, 'a consensus decision is reached involving everybody'. The client agent, too, agreed that it 'always [made] sure that everybody has a chance to put their views across'.

Situational versus prompted knowledge

The cognitive map shown in Figure 4.10 forms the basis for the following discussion.

The case study showed that the focus during change events was to solve the practical problem in hand rather than promote knowledge creation for future use. As the D&B contractor stressed they 'can come up with new ideas which are not practical'. The architect stated: 'we put together bits of pieces discussed at meetings through the drawings. This is information co-ordination rather than new idea generation.' There was some scope for creating new knowledge. For example, the team did trial-and-error work to find out solutions when the client requested them to do so.

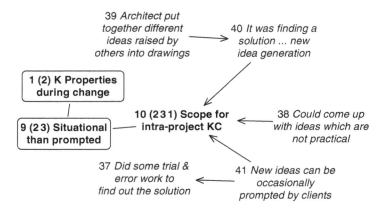

Figure 4.10   Project A: cognitive map of situational versus prompted knowledge.

Table 4.2   Project A: evidence relating to proposition P2.

Attributes	Evidence	Confirmation/falsification
Tacit than explicit	▪ Reliance on previous knowledge ▪ Limited chance to use explicit knowledge	Confirmed
Collective than individual	▪ Giving everybody a chance ▪ Dependency ▪ Consensus decision	Confirmed
Situational than prompted	▪ Limited scope for new ideas ▪ Solve practical problem	Confirmed

Comment on proposition P2

> P2: *The knowledge that the project team members use in the reactive change process is: more tacit rather than explicit; more collective (mutual) rather than individual; and more situational rather than prompted.*

The case study findings showed how team members relied on their past experience and how they shared and used this knowledge in solving the change problem. Further, the data confirmed that there was less opportunity and need to use explicit knowledge. Table 4.2 summarises these findings that support the claim that the team used tacit knowledge more often compared to explicit knowledge when they managed change. This confirms the first part of proposition P2, namely that the knowledge that the project team members use in the reactive change process is more tacit in nature than explicit.

The concepts such as giving everybody a chance, dependency between team members, and reaching a consensus decision suggest that the change event was solved collectively rather than by an individual team member. The second part of proposition P2, namely that the knowledge that the project team members use in the reactive change process is more collective (mutual) than individual, is confirmed.

The concepts such as limited scope for new ideas, and need to solve the practical problem in hand, suggest that the knowledge created during change events is situational rather than prompted; this thus confirms the third part of proposition P2, namely that the knowledge that the project team members use in the reactive change process is more situational than prompted.

Taking all three properties into account the case study confirms proposition P2, namely that the knowledge that the project team members use in the reactive change process is: more tacit than explicit; more collective (mutual) than individual; and more situational than prompted.

4.2.4   Knowledge identification and utilisation during change

This section presents the coding structure (Figure 4.11) and the cognitive map (Figure 4.12) that relate to the knowledge identification and utilisation aspect. These form the basis for the following discussion, while an overall summary and comment on proposition P3 are provided at the end.

The data revealed that the team was in a good position to identify where the knowledge within the team resided. This was due to prior relationships that the team had developed through the partnering arrangement. Some team members also had opportunity to work in parallel projects, which further strengthened their relationships. As the D&B contractor asserted:

> the partnering arrangement made good team relationships and exploited previous knowledge in working on client's projects.

In utilising relevant knowledge for managing the change event, different team members were involved in the process. As the architect stated, 'we are dealing with experts', therefore, the input of different members was required at different points. For example, the architect brought in technical knowledge, while the D&B contractor contributed general construction and programme knowledge. In addition, a significant feature was that the input of the flooring sub-contractor was utilised in this change event. As the client agent stated, 'the specialist sub-contractor's knowledge was valued and welcomed'. They brought practical knowledge to the issue.

General issues relating to this change were discussed at the regular design team meetings where everybody participated. However, the key members who were actively involved throughout the process were the client, the client agent, the contractor and the architect. These members held specific meetings outside regular meetings to discuss the issue in depth. As the client explained:

> there was no need for everybody to participate in these [specific meetings], for example mechanical-electrical has nothing to do with the flooring.

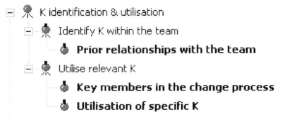

Figure 4.11 Project A: coding structure of knowledge identification and utilisation. Screen shot courtesy of QS International.

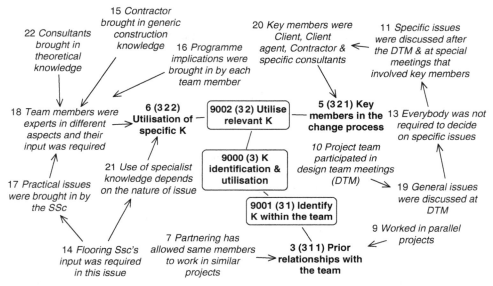

Figure 4.12 Project A: cognitive map of knowledge identification and utilisation. DTM, design team meeting; SSc, specialist sub-contractor.

Comment on proposition P3

> *P3: Project team members, who know where the knowledge resides within the team through prior interactions, are better at identifying and actively utilising relevant team members' knowledge, during the reactive change process, compared to project team members who do not know where the knowledge resides within the team.*

The team was able to utilise relevant team members' knowledge effectively during the change event as they knew where the knowledge resided within the team through the development of the partnering relationship. The involvement of different members; use of specialist sub-contractors' knowledge; and holding of specific meetings suggest that the relevant knowledge was utilised by the team (Table 4.3).

Table 4.3 Project A: evidence relating to proposition P3.

Attributes	Evidence	Confirmation /falsification
Identify and utilise relevant knowledge	▪ Different members involved ▪ Use of specialist sub-contractors' knowledge ▪ Specific meetings with key people	Confirmed

Figure 4.13 Project A: coding structure of intra-project knowledge creation. Screen shot courtesy of QS International.

4.2.5 Intra-project knowledge creation during change

Figure 4.13 presents the coding structure in relation to intra-project knowledge creation. The case study data are then discussed under socialisation, externalisation, combination and internalisation stages. Finally, an overall summary and the comments on proposition P4 and sub-propositions SP4.1 to SP4.4 are provided.

Socialisation stage

This section addresses the case study findings on the socialisation stage, along with the relevant cognitive map (Figure 4.14).

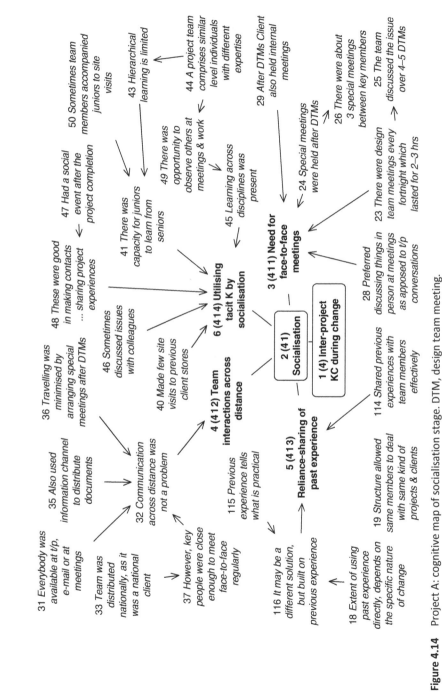

Figure 4.14 Project A: cognitive map of socialisation stage. DTM, design team meeting.

As discussed under knowledge properties, the team members effectively shared and relied on their past experience in managing the change event. The team had opportunities to create new tacit knowledge based on their existing tacit knowledge (socialisation mode), when they engaged in shared problem-solving through face-to-face settings.

The case study showed that the team members used face-to-face settings to a significant extent. For example, the architect said, 'mostly at meetings the issues were discussed in detail' and the D&B contractor replied, 'decisions were made mainly in these meetings'. They held design team meetings (DTMs) every fortnight, which generally lasted from 2 to 3 hours. In addition, the key people who were actively involved in the change event held their own meetings to discuss specific issues relating to the change event. The client agent had separate meetings with the client and with the D&B contractor, for example, to discuss cost and programme issues. The use of meetings was further evident when the client stated, 'after I had a meeting with the contractor I have another with the store'. The architect emphasised the importance of having meetings over other communication mediums: 'when using telephones we cannot discuss much. I prefer describing things by use of diagrams, in person at meetings.'

The team was geographically dispersed, as they dealt with a national client. The distance barrier, however, did not affect the communication between them as the key people were close enough to meet face-to-face regularly. Further, the travelling was minimised by arranging special meetings after regular design team meetings. As well as meetings, the team used electronic co-ordination (information channel) to distribute project documentation between parties.

In addition to developing team members' tacit knowledge through meetings, there were other forms of tacit knowledge utilisation. The team, for instance, made site visits to similar stores and participated at social events where they made close contacts with other team members. The team also had opportunities to observe and learn from their colleagues who were physically co-located in the workplace. Also, there was capacity for juniors to learn from senior members during site visits. Nevertheless, in this case study the tacit-to-tacit knowledge transfer was more evident between the same-level team members rather than vertical transfer from a senior to junior. In the D&B contractor's words:

> Hierarchical learning is limited as there tends to be similar level people involved. But learning across other disciplines is present.

The client agent, too, agreed to this by stating, 'at project meetings I had opportunity to observe other members and their work'.

Externalisation stage

This section addresses the case study findings on the externalisation stage, along with the relevant cognitive map (Figure 4.15).

It was evident that team members used different ways to externalise their tacit knowledge during discussions connected to the change event. They used

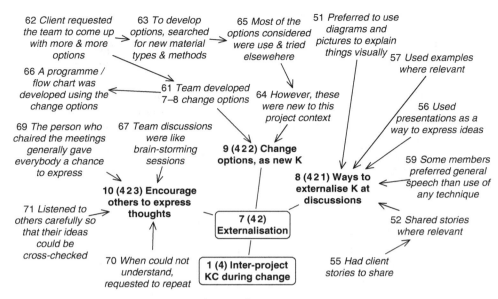

Figure 4.15 Project A: cognitive map of externalisation stage.

diagrams and pictures to explain things visually. They also used previous projects as examples to explain floor options and brought in stories of previous projects where relevant in discussions. The D&B contractor said that the presentation they made to the client helped in clarifying their thoughts. The client agent, however, said:

> I like to keep it as simple as possible and express in few words. I personally do not have a habit of using other ways to express things.

Thus, the use of different externalisation mechanisms depended on the individual personality and also the discipline. The architect, for instance, said that he always preferred to explain his ideas through a sketch or a drawing.

Team members encouraged each other to externalise their thoughts during the change decision-making process. Everybody was given a chance to express ideas and the discussions were more like brainstorming sessions. As the D&B contractor stated, 'there are kind of brainstorming sessions or open meetings that encourage people to come with better solutions.' The client agent said that he carefully listened to other members, so that he could cross-check others' ideas with his ideas. It was natural to ask a team member to repeat what was said when others did not understand. The team developed about seven to eight change options by externalising and discussing their ideas in this way.

Combination stage

This section addresses the case study findings on the combination stage, along with the relevant cognitive map (Figure 4.16).

The externalised tacit knowledge during team discussions was captured by the team in written forms at different levels. The minutes of meetings were one way that the team codified the externalised knowledge during team discussions. However, these minutes did not have sufficient detail to be of use in other contexts. The D&B contractor emphasised this when he stated:

> Meeting minutes are a way of bringing things that we discussed together. But a drawback is the discussions are not reported in detail at these minutes. We talk for about 40 minutes and include in the minutes what we agreed at last and not the pathway of reaching that decision.

The other codified documents were the change request forms. These documents, however, included only directions for implementing a particular change decision and not details of the total change process. The architect viewed 'revised drawings' as a form of codification of the change event. However, they again contained details of the agreed change rather than the total change process.

There was no evidence of a project review report at the completion of the project. The D&B contractor stated that 'no special reports were written after the meetings other than these minutes of meetings', and the architect affirmed, 'no formal project review reports were prepared'. However, some project feedback was received through a series of meetings that were held after project completion.

It was apparent that the change was recorded in an ad hoc manner. There was no particular party responsible for the recording of the change experience. Even though the client agent was supposed to maintain the CRF procedure, it was the D&B contractor who kept these records. Individual organisations kept their own records at different levels for different purposes. For example, the D&B contractor kept records for quality assurance purposes and financial implications, and the client kept records of the consultants to evaluate their performance.

Internalisation stage

This section addresses the case study findings on the internalisation stage, along with the relevant cognitive map (Figure 4.17).

In terms of internalising knowledge that was created during the change process, it was apparent that the team members gained tacit knowledge, to some extent, by going through the change experience. For example, the team learned about technical issues such as solutions for floor lifting problems, methods for testing poor floor conditions and methods of laying tiles during difficult circumstances. The architect confessed, 'in addition to learning how to deal with the new technical issues we have become more flexible and understanding of change.' The D&B contractor stated, 'after going through that experience and the problems we encountered, we come to a situation where we can handle

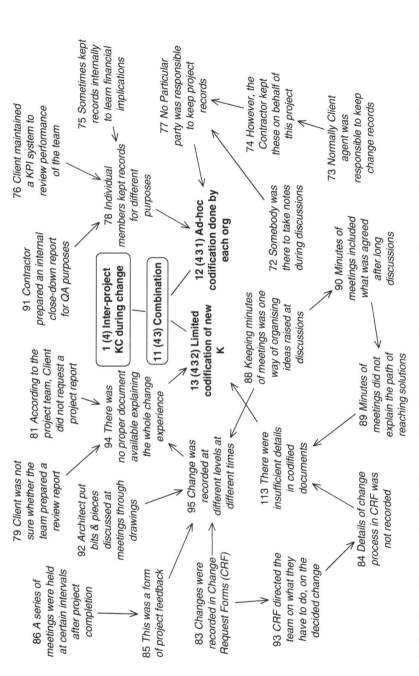

Figure 4.16 Project A: cognitive map of combination stage. CRF, change request form; KPI, Key Performance Indicators; QA, Quality Assurance.

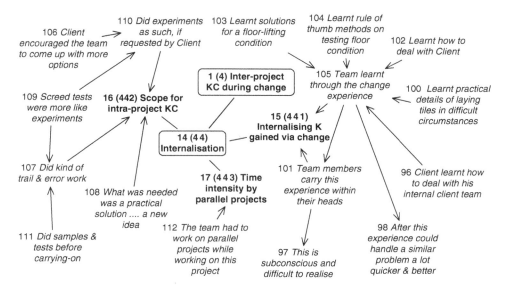

Figure 4.17   Project A: cognitive map of internalisation stage.

that problem a lot quicker and better.' Even the client admitted that he learnt 'probably to push harder higher up the line and try and get the decisions quicker.' The client agent stressed, 'I learnt how to deal with the client, how to make sure he understands and how long he needs to understand etc.' The team said that such learning is generally carried out subconsciously in their heads.

Team members, however, had limited opportunities to internalise and reflect upon their experience to any great extent. They were constrained by time and by having to work on parallel projects. There was little scope for experimentation and reflection. Yet some forms of experiment did take place such as screed tests, doing samples and coming up with new ideas in response to client requests. But the team did not have further opportunities to reflect deeply on their change experience afterwards and effectively feed-forward their learning in future projects.

Comment on proposition HP4 and sub-propositions SP4.1 to SP4.4

> *P4: Project teams are more likely to create new knowledge based on the existing knowledge during the reactive change process, through a natural flow of knowledge rather than through a full cycle of knowledge conversion process.*
>
> *SP4.1: Project teams who interact with other team members regularly through face-to-face settings, during the reactive change process, are better at utilising existing tacit knowledge and creating new tacit knowledge, compared to project teams who do not interact regularly through face-to-face settings.*
>
> *SP4.2: Project teams who actively use visualisation techniques during team discussions that arise during the reactive change process are better*

at expressing and externalising their tacit knowledge compared to project teams who do not use such techniques.

SP4.3: Project teams are unlikely to effectively combine and codify externalised tacit knowledge arising out of team discussions in reaction change situations.

SP4.4: Project teams are more likely to acquire superficial learning through the change experience, during the reactive change process, rather than effective internalisation through reflection.

This section first summarises the findings in relation to sub-propositions SP4.1 to SP4.4 and finally considers proposition P4.

Tacit-to-tacit knowledge transfer was enabled by socialisation activities between the team members. There were regular design team meetings, special meetings, site visits and social events (Table 4.4.). Thus, sub-proposition SP4.1 is confirmed; namely, project teams who interact with other team members regularly through face-to-face settings during the reactive change process, are better at utilising existing tacit knowledge and creating new tacit knowledge compared to project teams who do not interact regularly through face-to-face settings.

The team showed evidence of externalising their tacit knowledge to explicit knowledge during discussions through different techniques (see Table 4.4), especially visualisation techniques. Thus, sub-proposition SP4.2 is confirmed, namely, that project teams who actively use visualisation techniques during team discussions that arise during the reactive change process are better at expressing and externalising their tacit knowledge, compared to project teams who do not use such techniques.

As discussed above, the combination of externalised knowledge was not carried out to provide a detailed account and there was no particular party formally responsible for capturing the lessons learnt on change events. This was substantiated by evidence such as lack of details in minutes of meetings, drawings and change record forms; absence of project review reports; and ad hoc codification done by individual members (see Table 4.4). Thus, sub-proposition SP4.3 is confirmed, namely, project teams are unlikely to effectively combine and codify externalised tacit knowledge arising out of team discussions in reaction change situations.

The team learnt, to some extent, through activities such as resolving technical issues; dealing with the client; discovering how to handle such issues quicker and better; and becoming more flexible and understanding of change. However, evidence, such as the time pressures due to parallel projects, suggests that the team did not get the opportunity to extend their learning to deep internalisation through reflection (see Table 4.4). Thus, the case study findings confirm sub-proposition SP4.4, namely that project teams are more likely to acquire superficial learning through the change experience, during the reactive change process, rather than by effective internalisation through reflection.

On the whole, the sub-propositions SP4.1 to SP4.4 confirm that the team gained new knowledge through interactive settings while going through the change experience. However, a full cycle of knowledge conversion

Table 4.4   Project A: evidence relating to proposition P4.

Attributes	Evidence	Confirmation/falsification
Presence of socialisation activities	▪ Regular meetings ▪ Special meetings ▪ Site visits ▪ Social events ▪ Seating arrangement ▪ Observing other members ▪ Seniors accompanying juniors	Confirmed
Use of different techniques to externalise	▪ Diagrams and pictures ▪ Previous example projects ▪ Stories ▪ Presentations ▪ Brainstorming sessions ▪ Give everybody a chance ▪ Encourage others ▪ Listen to others	Confirmed
Limited combination	▪ Lack of details in minutes of meetings ▪ Recording only the final decision in CRF ▪ Limitation of drawings as a change record ▪ Absence of project review reports ▪ Ad hoc codification by individual parties	Confirmed
Superficial internalisation	▪ Learnt technical issues ▪ Learnt how to deal with client ▪ Became flexible and understanding of change ▪ Can handle a lot quicker and better next time ▪ Time intensity through parallel projects ▪ Limited experiments, such as tests and samples ▪ Limited client requests to consider new ideas	Confirmed

CRF, Change Request Form.

from socialisation to internalisation via externalisation and combination modes was not apparent. The team gained tacit knowledge through socialisation and internalisation, but failed to properly combine the externalised knowledge. Thus, proposition P4 is confirmed, namely that project teams are more likely to create new knowledge based on the existing knowledge during the reactive change process, through a natural flow of knowledge rather than through a full cycle of the knowledge conversion process.

Figure 4.18 Project A: coding structure of inter-project knowledge transfer. KT, Knowledge Transfer. Screen shot courtesy of QS International.

4.2.6   Inter-project knowledge transfer via change

Figure 4.18 provides the coding structure that relates to inter-project knowledge transfer. This is discussed in two parts: first, transfer of knowledge via individuals; and second, inter-project knowledge dissemination. Finally, a summary of the findings and comments on proposition P5 and sub-propositions SP5.1 to SP5.2 are provided.

Transfer of knowledge via individuals

The cognitive map that relates to 'transfer of knowledge through individuals' is given in Figure 4.19. This forms the basis for the following discussion.

Team members developed tacit knowledge through the change experience. The client agent described the difficulty in transferring tacit knowledge as follows:

> What seems to happen in projects is there's always a particular problem and the individuals involved go through the experience and learn the best solution for the given problem. But this does not pass to other members. This same flooring problem may be experienced at the same time by a colleague, without my knowledge … This is the problem. Unfortunately, it is only the individuals involved who carry this knowledge in their heads.

As discussed under knowledge properties, the case study also revealed that there were few opportunities to re-use explicit knowledge that was created through project activities. Team members admitted that documents were limited in giving direct solutions to practical issues, as it is generally a new situation that they are having to deal with. Further, the project documents could be useful only if the team members could recall a specific thing from their previous project experiences.

It was evident that team members had opportunities to apply knowledge that they gained through this change event in other parallel projects that they were involved in. For example, the architect said, 'same flooring problem was experienced in another project, but the solution considered finally was different', and the D&B contractor noted, 'we had a job running eight weeks later which had the identical floor. Based on the experience we had going through this project we decided not to consider terrazzo flooring replacement for that project.' But the client agent said that he did not get an opportunity to re-use this specific knowledge, as he did not get a chance to work in the same type of retail projects afterwards.

In terms of using knowledge generated from the flooring project in future projects, team members were positive that their level of understanding had improved because of the experience and that they would be more flexible effectively. That being said, the client agent said, 'the process could be made efficient and effective next time, only if the client makes prompt decisions. So as you see this depends on individuals involved.'

Inter-project knowledge dissemination

The cognitive map that relates to inter-project knowledge dissemination is given in Figure 4.20. This forms the basis for the following discussion.

In terms of disseminating knowledge gained through a project change experience at the multiple organisation level, it was clearly evident that face-to-face settings were widely used. There were monthly meetings within the organisations where the employees reported and discussed their project progress with a wider audience; and there were annual conferences where the progress of the firm as a whole was reported. The employees also had opportunities to share their different project experiences with other colleagues, through face-to-face settings. The architect stated, 'there are two architects from our firm who deal with client projects. We are seated close to each other, so do share certain memories together', and the client agent stated, 'the open-plan layout that we have in this new office premises provides more interactions with the staff'. In addition, the firms had links with other organisations. For example, the D&B contractor worked with an alliance of contractors, where they held meetings once a month to disseminate best practice, and the client agent said, 'as individuals we are members of different external organisations and committees ... we do participate in social activities organised by these'.

The use of meetings as a mechanism to disseminate project knowledge was further evident when the project team held a series of project review meetings after project completion. The team also held a social function on

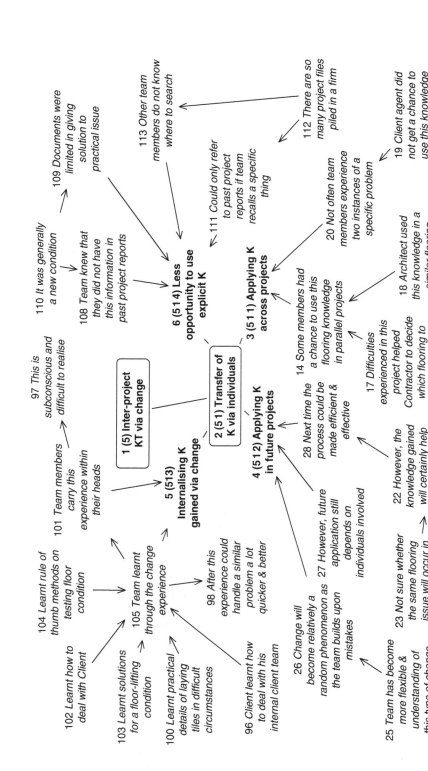

Figure 4.19 Project A: cognitive map of transfer of knowledge via individuals.

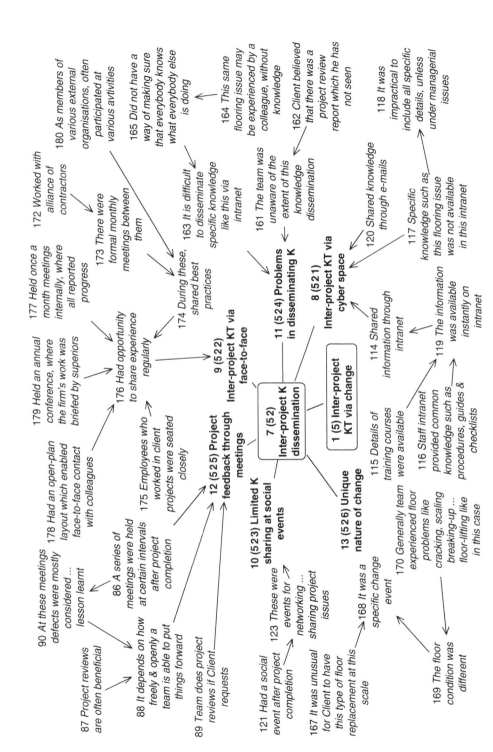

Figure 4.20 Project A: cognitive map of inter-project knowledge dissemination.

the project completion. However, as the architect and the client agent stated, they were networking events rather than events for sharing technical experience.

In addition to these face-to-face settings, employees also shared knowledge through cyberspace. For example, they used e-mails to share knowledge, as the client agent explained, 'it is common to send an e-mail by an individual to the rest asking a certain question'. They also had access to intranets, where common knowledge was available instantly. But these did not provide specific knowledge on issues such as the flooring problem. The team members indicated the difficulty of disseminating such specific knowledge via intranets. Thus, the specificity of newly created knowledge is an influencing factor on the extent of knowledge dissemination. The client agent further emphasised the problem of knowledge dissemination:

> This same flooring problem may be experienced at the same time by a colleague, without my knowledge. We do not have a way of making sure that everybody knows what everybody else is doing.

Team members were not certain to what extent knowledge was disseminated within their individual firms.

Comment on proposition P5 and sub-propositions SP5.1 and SP5.2

> *P5: The new knowledge that is created during the reactive change process is transferred to multiple organisations, for potential re-use in future projects, through personalisation strategies rather than through codification strategies.*
>
> *SP5.1: The new knowledge created during the reactive change process is re-used in future projects through the individuals involved during the process, rather than through the codified documents.*
>
> *SP5.2: The new knowledge created during the reactive change process is disseminated and made available to the wider organisation for potential future re-use, through interactive settings between the organisational members rather than through effective dissemination of codified documents.*

This section first summarises the findings in relation to sub-propositions SP 5.1 and SP 5.2 and finally considers proposition P5.

The findings show that knowledge gained through a change experience is mainly passed on to a similar context through the individuals involved. Thus, sub-proposition SP5.1 is confirmed; namely, the new knowledge created during the reactive change process is re-used in future projects through the individuals involved during the process rather than through the codified documents.

At the firm level there are limited opportunities for disseminating knowledge through codification strategies. The knowledge is disseminated informally through various face-to-face settings. This is substantiated by evidence

Table 4.5   Project A: evidence relating to proposition P5.

Attributes	Evidence	Confirmation/ falsification
Transfer to future projects via individuals	▪ Less chance to re-use explicit knowledge ▪ Members applying knowledge in other projects	Confirmed
Dissemination via personalisation	▪ Monthly meetings ▪ Annual conferences ▪ Seated closely/open-plan layout ▪ Meetings with other firms ▪ Project review meetings ▪ Social functions ▪ Absence of specific knowledge in intranet	Confirmed

of face-to-face interactions at the firm level such as monthly meetings, annual conferences, and project review meetings (Table 4.5). Sub-proposition SP5.2 is confirmed; namely, the new knowledge created during the reactive change process is disseminated and made available to the wider organisation for potential future re-use through interactive settings between the organisational members, rather than through effective dissemination of codified documents.

Considering sub-propositions SP5.1 and SP5.2, the case study indicates that new knowledge created during change events is passed to future projects through personalisation strategies rather than codification strategies. Proposition P5 is therefore confirmed; namely, the new knowledge that is created during the reactive change process is transferred to multiple organisations, for potential re-use in future projects, through personalisation strategies rather than through codification strategies.

4.3   Case Study: Project B

4.3.1   Case study description

The case study project was a secondary school building. The project used a D&B procurement path. The project's duration was 15 months and its value £8.25 million. The change events considered were 'change of height in white boards' and 'change of design in radiant panels'. The D&B contractor, the architect, and the M&E (Mechanical and Electrical) consultant were interviewed for the two change cases. Taking into account the similar project situations, these two change cases are combined and analysed in this section.

In the whiteboard change case, the whiteboards in the classrooms were already fixed to the walls by the contractors when the client party noticed that they were too high for teachers to reach. This height issue had already been raised at initial meetings but there had been a significant change in the membership of the in-house D&B management team during this period. The old team had failed to put in writing what was discussed and agreed. As a result the new team were not aware of what had been agreed before. The new team, therefore, proceeded with constructing the whiteboards as per the drawings, which they believed had been approved by the client. The client party had also failed to pick up the incorrect specification when the new team informed them that the whiteboards would be fixed as per the drawings. The client insisted that the whiteboards be lowered. As a result, the contractor had to re-fix the whiteboards along with all of the associated work including wiring, painting, plastering and cleaning. The construction side of this issue took two weeks to resolve, but the financial aspects were not sorted out until the completion stage of the whole project.

In the radiant panel case, the M&E consultant had made a design error with respect to the positioning of the radiant panels within the classrooms. The centre of the room was not getting sufficient heat as per the original design. When the issue was noted by a specialist in the client party the design had to be changed. This change did not affect the project progress to a great extent as the error was detected and corrected prior to the installation of the radiant panels.

4.3.2   Contextual factors of change

Contextual factors of change are discussed in this section under process characteristics, group characteristics and organisational characteristics. Finally, a summary of the findings relating to contextual factors and comments on proposition P1 are given.

Process characteristics

In this section the coding structure (Figure 4.21) and the cognitive map (Figure 4.22) that relate to process characteristics are presented. These form the basis for the following discussion, which is described under two headings: 'design-driven triggers' and 'task dependency'.

Design-driven triggers
The incorrect positioning of the whiteboards was realised after they had been installed. A survey of the teachers had been conducted to ascertain their average height to inform the positioning of the whiteboards. When a teacher visited the site he noted that the whiteboards did not satisfy this average height. The cause of this change was triggered from the design stage due to miscommunication between parties. When the old in-house D&B team left the project in the middle of this change case, they failed to write down what was agreed between them and the client at previous meetings. In the D&B contractor's words this was caused by:

> ... bad communication really. No one ever closed the issue. People were obviously thinking about it at some stage. It had been mentioned at meetings before we got involved at this site. But no one had actually put this in writing that this is what is required.

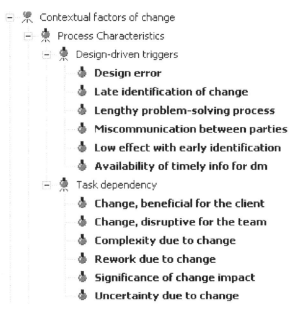

☐ 🐾 Contextual factors of change
  ☐ 🐾 Process Characteristics
    ☐ 🐾 Design-driven triggers
      🐾 **Design error**
      🐾 **Late identification of change**
      🐾 **Lengthy problem-solving process**
      🐾 **Miscommunication between parties**
      🐾 **Low effect with early identification**
      🐾 **Availability of timely info for dm**
    ☐ 🐾 Task dependency
      🐾 **Change, beneficial for the client**
      🐾 **Change, disruptive for the team**
      🐾 **Complexity due to change**
      🐾 **Rework due to change**
      🐾 **Significance of change impact**
      🐾 **Uncertainty due to change**

Figure 4.21   Project B: coding structure of process characteristics. Screen shot courtesy of QS International.

The new D&B team, therefore, installed the whiteboards as per the height given in the design stage drawings. The client party, too, failed to inform the new team that it needed to reduce the height of the whiteboard from the original design. The architect said, 'we believed that we got the details from the contractor and that the client had fed the criteria to the contractor', and the M&E consultant said, 'the architect issued the drawings with the dimensions on. We assumed that the dimensions given were the correct dimensions and agreed with the client'.

The whiteboard change took some time to resolve. The construction issue was resolved within two weeks but the financial issue persisted until the completion stage of the project. The D&B contractor stressed that:

> The actual decision-making process was more time consuming than complicated. As you know we were not at site from the start. All the correspondence took place with the client 6 months prior to our involvement. So I spent couple of days going back to previous minutes of meetings and trying to find out what went wrong. So more time consuming in that way.

The reason for the radiant panel case was, again, triggered by a design error. The D&B contractor explained, 'in the original M&E design it was positioned on the wall. This was queried by the client's specialist. So the M&E designer checked again and he realised that the heating in certain spots would not work and that is how it was identified.' Even though the

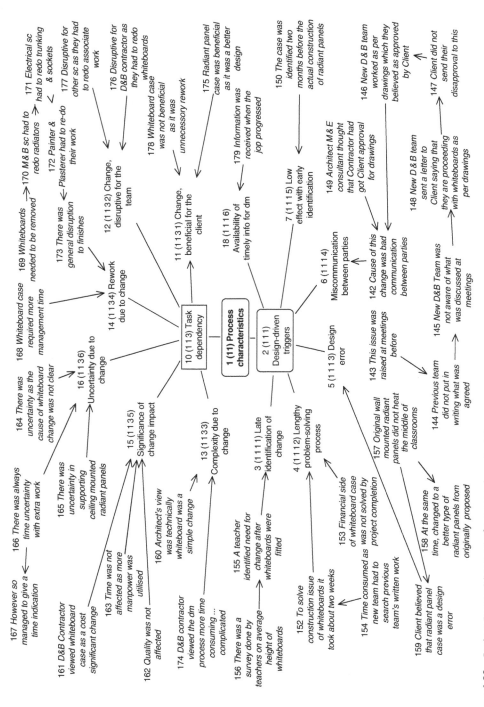

Figure 4.22 Project B: cognitive map of process characteristics.

solution was straightforward (to change from a wall mounting to a ceiling mounting), it still took two weeks to resolve. The D&B contractor stated, 'it took about two weeks to work up the new design. The actual time is about a day and a half, but it took two weeks to resolve.' The early identification of this change, however, reduced possible disruptive effects and it was considered beneficial by the client.

Task dependency
The whiteboard case affected a wide range of parties and caused rework. It was disruptive to both the client's side and the contractor's side. From the D&B contractor's viewpoint change was, 'completely not beneficial. Obviously after the change now everything works. But it should have been done right first time.'

The extent of rework caused by the change was described by the D&B contractor: 'we had to put more management time to sort this out … It wasn't quite simple – just dropping the whiteboard. The radiator, plug sockets and trunking had to be moved.' The sub-contractors, therefore, were affected by this change. For example, the D&B contractor explained the mechanical and electrical sub-contractor's involvement:

> … he had to move the radiators. He worked two weekends. So they were on over-time rates, which we have to pay. They got about six men.

He further added:

> The painter and the plasterer had to redo the work and then there was general disruption to finished work. The rooms were complete and were locked up. So after re-doing the plaster and painting we had to go back and clean up.

Thus, the change affected many parties and created rework.

The whiteboard change was viewed by the D&B contractor as costly. The architect expressed a different opinion, saying that it was a small change as it was not technically complex. He stated:

> I think this is relatively speaking a fairly small change from our perspective. If you were talking about changing steel frame or a concrete frame or a load-bearing wall or changing a floor construction I think I will be more involved in this issue.

The overall quality and duration of the project was not affected as the team tried to meet the same quality standards and utilised more people to meet the set time targets. The D&B contractor described this, 'because we resourced it with extra people it did not affect the project time … quality-wise it shouldn't have made a difference.' There was uncertainty created by the change. The architect explained this, 'there was very much uncertainty because we were not clear how this error happened in the first place. I had to go back and review the files to get my thoughts on that', and the D&B

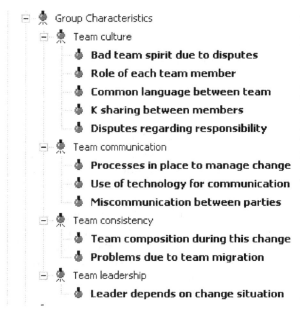

Figure 4.23   Project B: coding structure of group characteristics. Screen shot courtesy of QS International.

contractor expressed, 'there's uncertainty with any extra work. You never know how long it is going to take, etc. but we got an indication of time from the subcontractors as to the programmes'.

Group characteristics

The coding structure (Figure 4.23) and the cognitive map (Figure 4.24) that relate to group characteristics are presented in this section. These form the basis for the following discussion, which is described under four headings: team culture, team communication, team consistency and team leadership.

Team culture
The project team members had to rely on one another and perform different roles in managing these change events. The D&B contractor, for example, had a strong management role in the whiteboard case, while the M&E consultant had a pivotal role in the radiant panel case. The architect had to issue revised drawings. The client's role was to agree on the final solution. However, bad team spirits that existed between the parties impacted effective functioning of these roles towards effective change management. This was clearly evident in the whiteboard case when the team could not finalise the financial implications between the parties until the project completion. The main parties, the client and the contractor, argued over this matter and were not prepared to work together to resolve it. The D&B contractor described,

'we argued a lot on this – whose fault it was and who [was] going to pay for it, etc. That is still going on.' Hence, there was not a significant team bond between them. The D&B contractor explained the reason, 'it is because the job did not go very well. I think when things go wrong, the relationships tend to break.' The knowledge was shared openly and freely only between some team members due to these broken relationships. For example, the internal D&B team benefited from pooling knowledge and the knowledge was shared openly between the architect and the D&B contractor. The client and the contractor worked as two parties and were not sharing knowledge effectively. The architect put it across this way:

> If you put the client aside, the team I see is primarily the contractor and the design team, that team I think shared information fairly openly and freely.

The D&B contractor echoed this:

> I tried to always share, but not with the client. We have to be careful when we share with the client until we are certain of what we say really. As regard to ourselves and the design team, I am completely open.

When questioned whether or not specialist language sometimes hampered effective communication, team members said that it depended on the nature of the issue. For example, since the whiteboard case involved general construction knowledge the team did not have a difficulty. But the radiant panel case was found to be quite technical to understand by the D&B contractor.

Team communication
Communication between the parties was unsatisfactory due to ongoing disagreements and disputes. The team members did not show evidence of using technologies effectively to share knowledge other than the use of occasional e-mails and telephones to communicate. As the architect stated:

> Some contracts are set up with an IT facility so that we can exchange information electronically within the contract. The main way of exchanging information in this project was by way of hard copies.

However, as explained above, the team widely used face-to-face meetings as a mechanism to exchange knowledge.

A formal change management procedure was absent in this project. The change was communicated between parties through an instruction from the client. The D&B contractor stated, 'the process was a letter from the client telling us to proceed with re-doing the work. It did not need to be any more formal than that.' The architect expressed his unhappiness on how the D&B team handled the whiteboard change. In his words, 'I do not think in this project the change management process was a descriptive one.' He also expressed how he viewed project change:

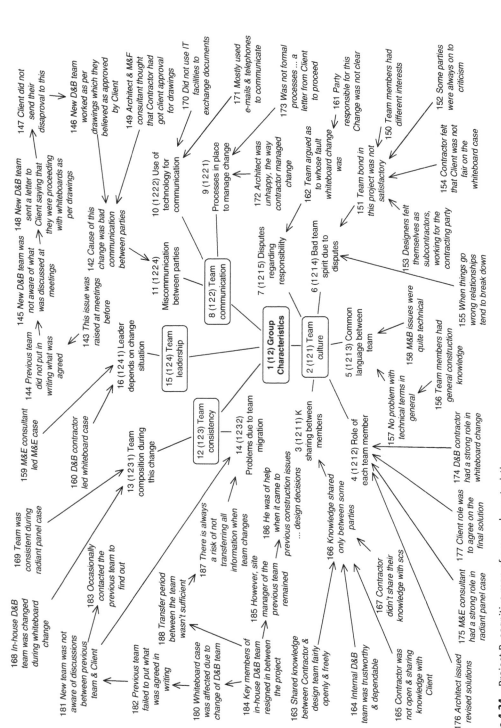

Figure 4.24 Project B: cognitive map of group characteristics.

> ... every project has changes and that is the biggest task for the project –
> to control changes that come in the construction process ... I would
> call it variations, construction variations, whether it generates from
> client variations, design team variations, variations from conditions of
> the project.

This suggests that for construction professionals a project change management process is basically maintaining a change order system.

Team consistency
The team composition during the whiteboard change was affected due to a large change in the D&B in-house team. The change in team membership created a series of problems and adversely affected the management of the change. The new team did not get enough opportunity to learn about the past phases of the project from the old team due to an insufficient transfer period. The architect declared:

> Obviously in any process when you have such a transfer period of staff
> resources there is a risk that the information may not have been trans-
> ferred ... the overlap between the new team coming in and the old team
> going out was not how it should be. It was not sufficient enough.

When D&B contractor was queried about whether he had contacted the old team after they left the project, he said that their project manager had contacted them on only a few occasions, and on those occasions there was not a productive exchange of knowledge. Also, the site manager who remained from the old team was not of significant help as he was not involved in the whiteboard case from the start. The D&B contractor explained this:

> The decisions had obviously taken place a long time before the actual
> construction started ... so he got more involved in the day-to-day con-
> struction work than in the design decision-making really. He was of
> help when it came to construction issues.

Team leadership
In terms of leadership during change events, it was evident that the party who led the change cases was very much driven by the nature of the issue. For example, the D&B contractor led the whiteboard case while the M&E consultant led the radiant panel case.

Organisational characteristics

The coding structure (Figure 4.25) and the cognitive map (Figure 4.26) that relate to organisational characteristics are presented in this section. These form the basis for the following discussion.

Figure 4.25 Project B: coding structure of organisational characteristics. Screen shot courtesy of QS International.

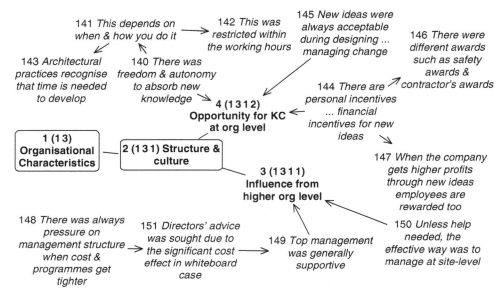

Figure 4.26 Project B: cognitive map of organisational characteristics.

Structure and culture

The case study data revealed that there was little influence from the structure and culture of the multiple organisations at a project level. The top management got involved only when it came to significant change events that involved high cost and time implications, such as the whiteboard case. For example, the D&B contractor stated, 'we discuss things when it comes to that kind of money. But in general, they [top management] support the site team in their decisions', and he further added, 'unless you need help, I think the right way to do this is at site-level really'.

In terms of creating new knowledge at the multiple organisation level, few opportunities existed. Even though there was autonomy to absorb new knowledge, this was constrained by work pressures. The architect described this as follows:

You always have freedom to absorb new knowledge, but it depends on whether you do that within your working hours or outside working hours. But you always have that option available to you.

The architect further added that the nature of his job provided capacity to consider new ideas and develop. For example he said, 'We always recognise CPD [continuous professional development] – you know, that the architectural profession recognises that you'll be given time to develop.' The firms did not encourage the employees by rewarding them financially for generating new ideas. According to the architect, 'there are no such rewards. I mean not financial incentives. The incentives might be personal incentives – your position in the company.' The D&B contractor affirmed this: 'as regard to financial incentives, no'. The M&E consultant explained his company's position:

> They value new ideas very much. Incentives and rewards are when the company gets higher profits the employees are rewarded. It is the profit that is important for the company.

Comment on proposition P1

> P1: *The reactive change process is very complex and ill-structured rather than simple and well-structured, due to various contextual factors that could be identified under process characteristics, group characteristics, organisational characteristics and wider-environmental characteristics.*

This section first summarises the findings in relation to process, group and organisational characteristics and then considers proposition P1.

Key process characteristics that affected these changes were triggered from the design phase. The whiteboard change was caused by miscommunication between parties, and the radiant panel case was caused by a design error. In terms of task dependency, the uncertainty created by the change disrupted the project members and incurred rework. It was the process complexity rather than the technical complexity that affected this change. These findings (Table 4.6) show the significant impact on project change from variables connected to design-driven triggers and task dependencies that could be grouped under process characteristics.

The case study findings suggested that financial implications of changes had influenced a team culture that was characterised by poor team spirit and an ineffective knowledge-sharing environment. Team communication was affected by the disputes that arose between the main parties. In addition, the team communicated and exchanged information during change by use of face-to-face settings rather than by the use of IT. Team inconsistency in terms of different representatives at project meetings had adversely affected the change decision-making process. Further, the nature of change had affected team leadership during a change event (see Table 4.6). These findings show the complex influence on project change from variables connected to group characteristics.

Empirical data further suggested that the organisation structure and culture did not have a direct influence on the project-level changes. The top management got involved when it came to significant changes in terms of cost and time pressures. The learning and knowledge-creating culture at the organisation level was not motivated by explicit rewards (see Table 4.6).

Table 4.6   Project B: evidence relating to proposition P1.

Attributes	Evidence	Confirmation/falsification
Process characteristics	▪ Design-driven triggers such as errors and miscommunication between parties ▪ Task dependencies such as uncertainty and process complexity	
Group characteristics	▪ Team culture with ineffective knowledge-sharing and poor team-working ▪ Team communication with regular meetings ▪ Team inconsistency with different representatives ▪ Team leadership that depended on the issue	Confirmed
Organisation characteristics	▪ Flexible organisation structure ▪ Ineffective learning organisation culture	

Considering these findings as a whole, it is difficult to establish clear relationships between process, group and organisational characteristics as these factors had a complex influence on project change. While certain factors positively impacted on project change, some factors incurred a negative impact. Thus, the degree and direction of the impact varied. Therefore, these findings confirm proposition P1; namely, the reactive change process is very complex and ill-structured rather than simple and well-structured, due to various contextual factors that could be identified under process characteristics, group characteristics and organisational characteristics. The case study, however, failed to provide sufficient evidence for wider-environmental characteristics.

4.3.3   Knowledge properties during change

Figure 4.27 presents the coding structure in relation to 'knowledge properties'. This is discussed in three parts below. The final part of this section provides the overall summary and the comment on proposition P2.

Tacit versus explicit knowledge

The cognitive map (Figure 4.28) that relates to 'tacit versus explicit knowledge' is presented in this section, along with a discussion of the case study findings.

The team members relied on their past experience in finding solutions to the two change cases of this project. The team referred to documents in

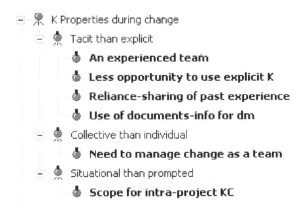

Figure 4.27   Project B: coding structure of knowledge properties. Screen shot courtesy of QS International.

order to find out the extent of the problem, rather than exploring novel options to solve the problem.

The team had general construction experience and specific experience on some school projects. The D&B procurement arrangement allowed the team to work collaboratively under one party, with the exception of the client. However, due to the absence of a partnering arrangement the team did not have much working experience with other members of the team. However, the arrangement did provide the opportunity to experience different working practices within firms. According to the architect, 'I realise that each organisation has different strengths and weaknesses.' The team members used and shared this range of abilities and experience in this school project. As the D&B contractor described, construction project teams learn from failures, rather than successes, of previous projects. In the M&E consultant's words, 'you always bring up previous experience to discuss issues.' The tacit knowledge gained from previous experience was used subconsciously in solving the problem. The D&B contractor confirmed this point when he stated, 'these options just came to my mind based on the practical situation … that is subconscious'.

When queried to what extent the team used explicit knowledge, the M&E consultant explained that since a solution to a particular problem is dependent on specific project conditions, there were limited opportunities to use codified knowledge in other projects. His view was that it is difficult, and may, indeed, not be beneficial, to document very specific instances like these. However, team members did refer to documents of past phases of this project including the drawings, specifications and minutes of meetings.

Collective versus individual knowledge

The cognitive map (Figure 4.29) that relates to 'collective versus individual knowledge' is presented in this section. This forms the basis for the following discussion.

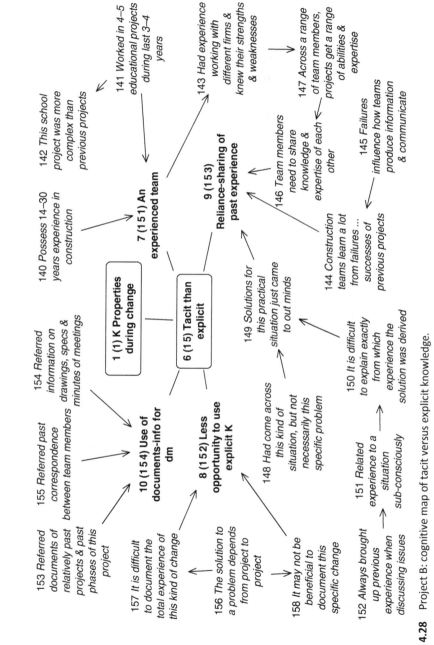

Figure 4.28 Project B: cognitive map of tacit versus explicit knowledge.

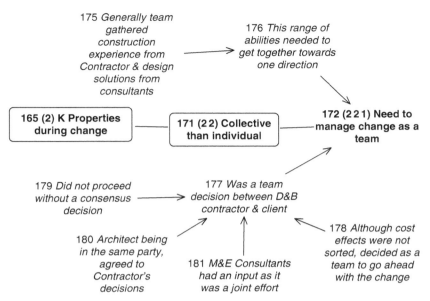

Figure 4.29   Project B: cognitive map of collective versus individual knowledge.

The team members described how they had to manage change collectively in the project. For example, the D&B contractor stated, 'it was a team decision between the client and us. We decided to do the work first and solve the money issue later.' The M&E consultant confirmed that it was a joint effort. The architect explained this collective effort further: 'generally, you gather construction experience from the contractor and get knowledge, ideas, solutions from the consultants and gel them into one building.'

Situational versus prompted knowledge

The cognitive map (Figure 4.30) that relates to 'situational versus prompted knowledge' is presented in this section, leading to a discussion of the case study findings.

The case study showed that the focus during change events was to solve, in a pragmatic way, the problem in hand using tried and tested knowledge, rather than to generate new ideas and promote knowledge creation. In fact, there was only one practical alternative available in the radiant panel case. In the M&E consultant's words, 'we had only one option. Because the heating was not reaching the middle of the room the only option was to mount from ceiling.' This shows the need to find a practical solution for a given situation rather than experiment with innovative ideas. From the architect's viewpoint, however, there was some scope for knowledge creation when designing. This was limited by cost and time pressures. The architect stated, 'with any design we face, we try to be innovative as much as we can within the design time and cost available.'

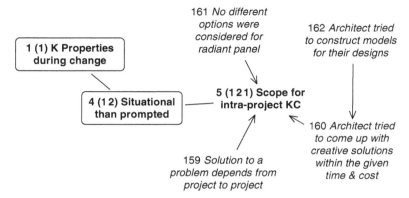

Figure 4.30   Project B: cognitive map of situational versus prompted knowledge.

Table 4.7   Project B: evidence relating to proposition P2.

Attributes	Evidence	Confirmation/falsification
Tacit than explicit	▪ Use and sharing of past experience ▪ Less opportunity to use explicit knowledge	Confirmed
Collective than individual	▪ Team decision ▪ Joint effort ▪ Use of different members' knowledge	Confirmed
Situational than prompted	▪ Limited practical solutions ▪ Limitation of new ideas generation	Confirmed

Comment on proposition P2

> P2: *The knowledge that the project team members use in the reactive change process is: more tacit rather than explicit; more collective (mutual) rather than individual; and more situational rather than prompted.*

The case study findings show that there was widespread use and sharing of tacit knowledge and experience but fewer opportunities to use explicit knowledge in generating and evaluating change options. This supports the first part of proposition P2; namely, that the knowledge that the project team members use in the reactive change process is more tacit in nature than explicit.

The concepts such as making a team decision, joint effort and use of different members' knowledge (Table 4.7) indicated that the change event was solved collectively rather than by any single, individual team member. This confirms the second part of the proposition P2, namely, that the knowledge that the project team members use in the reactive change process is more collective (mutual) than individual.

The concepts such as availability of limited practical solutions and limitation of new idea generation due to cost and time pressures (see Table 4.7) suggest that the knowledge created during change events was situational rather than prompted. This confirms the third part of proposition P2, namely, that the knowledge that the project team members use in the reactive change process is more situational than prompted.

Taking all these variables into account the case study confirms proposition P2, namely, that the knowledge that the project team members use in the reactive change process is: more tacit than explicit; more collective (mutual) than individual; and more situational than prompted.

4.3.4   Knowledge identification and utilisation during change

This section presents the coding structure (Figure 4.31) and the cognitive map (Figure 4.32) that relate to the knowledge identification and utilisation aspect. These form the basis for the following discussion, while an overall summary and comment on proposition P3 are given at the end.

The team members were unable effectively to identify and utilise knowledge within the team during the whiteboard case. For example, the team failed to acquire the end-users' requirements (teachers, in this case) before installing the whiteboards. Both the client and the contractor were confused as to why and how this issue arose. The two parties were blaming one another and, therefore, the financial side of this issue went on for a lengthy period without a solution. This was mainly due to the poor relationship between the client and the D&B team. Even though some of these team members had worked together before, none of the members had worked with the client. The architect explained the importance of prior relationships in managing change situations:

> It is interesting to know how much personal relationships affect your decision-making. Sometimes you tend to be flexible depending on the prior relationships.

In the radiant panel case, the knowledge of the M&E consultant was identified and utilised due to the specialised nature of the issue. The D&B contractor explained, 'I did not have expertise to evaluate design. It is quite complex as it involved heating calculations.' Thus, the use of specific knowledge depends on the nature of the issue and team members' functional roles.

Comment on proposition P3

> P3: Project team members who know where the knowledge resides within the team through prior interactions are better at identifying and actively utilising relevant team members' knowledge, during the reactive change

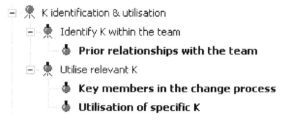

Figure 4.31 Project B: coding structure of knowledge identification and utilisation. Screen shot courtesy of QS International.

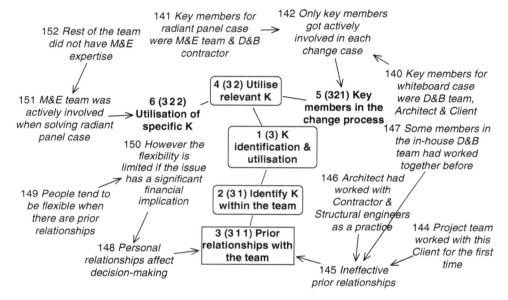

Figure 4.32 Project B: cognitive map of knowledge identification and utilisation.

process, compared to project team members who do not know where the knowledge resides within the team.

In summary, the whiteboard case showed that poor prior relationships affected the identification and use of knowledge that resided in the team. The concepts such as inability to consider end-user perspective, inflexibility of client decision and unresolved financial matters substantiated this claim (Table 4.8).

4.3.5   Intra-project knowledge creation during change

Figure 4.33 presents the coding structure in relation to intra-project knowledge creation. The case study data are then discussed under the socialisation, externalisation, combination and internalisation stages. The final

Table 4.8   Project B: evidence relating to proposition P3.

Attributes	Evidence	Confirmation/falsification
Failure to identify and utilise relevant knowledge	▪ Inability to consider end-user perspective ▪ Inflexibility of client decision ▪ Unresolved financial matters	Confirmed

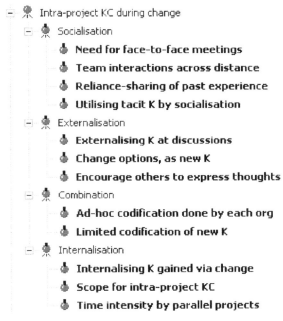

Figure 4.33   Project B: coding structure of intra-project knowledge creation. Screen shot courtesy of QS International.

part of this section provides the overall summary and the comments on proposition P4 and sub-propositions SP4.1 to SP4.4.

Socialisation stage

This section addresses the case study findings on the socialisation stage, along with the relevant cognitive map (Figure 4.34).

As discussed above, the team members shared and relied on their past experience to solve change cases. The team was thus able to gain new tacit knowledge based on their previous project experience (socialisation mode) through interactive settings that emerged during change events.

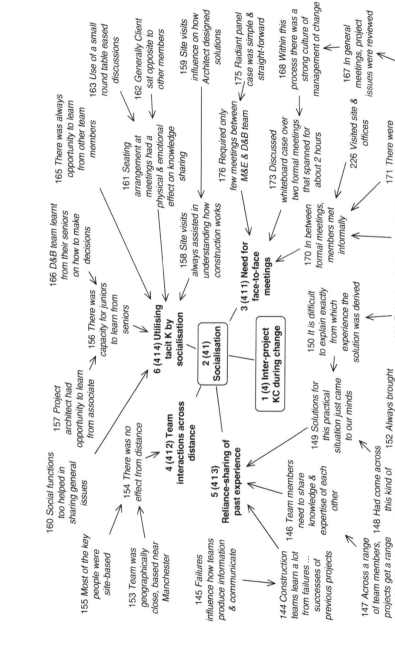

Figure 4.34 Project B: cognitive map of socialisation stage.

The empirical data revealed that the team members used face-to-face settings to a significant extent. The team held formal monthly meetings where they discussed project issues in general and also specific change events. The architect described this:

> Generally at regular meetings we review the main processes by looking at project issues, design issues and evaluation of events. Within that there was particularly a strong culture of management of change.

The M&E consultant confirmed this: 'the issues were taken up at several design team meetings and discussed.' In addition, there were informal meetings involving key people to discuss specific issues relating to change events. The D&B contractor affirmed this when he stated, 'it [whiteboard case] was discussed at normal progress meetings. But decided in separate site meetings without everyone there.' Even in the radiant panel case, which was a simple and straightforward change, the team held few meetings between the parties. The M&E consultant said, 'we had informal meetings with the D&B contractor'. The architect, too, highlighted these informal forums, 'between these major site meetings ... we meet the team directly at offices and site visits.' The D&B contractor further explained how they held separate meetings involving the in-house team:

> ... [we met] just going to the next office and sitting down for a cup of coffee. That goes sometimes for three hours or whole day ... It is the best way to do it really.

Since the team members were geographically close it was convenient to have face-to-face communication between them.

The team members had further opportunities to utilise tacit knowledge through other interactive settings. For example, the project architect had the opportunity to learn from the associate architect, who closely supervised him during this project; and the D&B team learnt from their top management on how to deal with the whiteboard case. The co-ordinator of the D&B team noted that he learnt from his project manager, who was more experienced than him. He stated: 'the project manager is more experienced than the rest of us'. He further said, 'I think you always learn something when you involve with people.' There was both horizontal and vertical tacit knowledge transfer between the team members. People also expressed the importance of site visits. The M&E consultant mentioned that the 'easy way to express your experience is to show others your work, for example take them to a site that shows the radiant panels installed'. The architect confirmed this when he stated, 'site visits always reinforce ... the more we go there the better'. Participating in social functions, too, assisted the team to share generic issues. Another interesting view that emerged from the empirical data was how seating arrangements at meetings could affect tacit

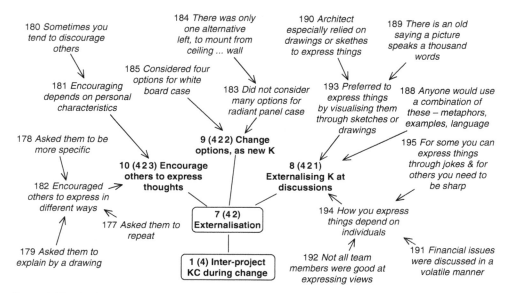

Figure 4.35   Project B: cognitive map of externalisation stage.

knowledge sharing. The architect said that, 'the table set-out has physical and emotional effect on the discussions really.'

Externalisation stage

This section addresses the case study findings on the externalisation stage, along with the relevant cognitive map (Figure 4.35).

The case study revealed that the team used different techniques to externalise their tacit knowledge during discussions that evolved around the change cases. The team members were very keen to use sketches and drawings to visualise their thoughts. The architect emphasised this when he was reminded of the old saying 'a picture speaks a thousand words'. When questioned about other techniques, he said, 'anyone would use all of those ... yes a combination of examples, metaphors and use of language'. It was evident that how team members expressed things at the discussions was dependent on individual characteristics. For example, the M&E consultant said, 'for some you need to say it by jokes to get something out and for another you may need to be sharp. So it depends on the individuals.' This also depends on the nature of the issue. According to the D&B contractor, the financial matters arising out of the whiteboard case were discussed in quite a volatile manner.

The team explained how they encouraged each other to externalise their thoughts and generate change options. For example, they asked others to be more specific or explain by a drawing or to repeat what was said. However, the architect raised another aspect, namely that depending on the type of person, on some occasions individuals would tend to discourage others rather than encourage them to express their thoughts.

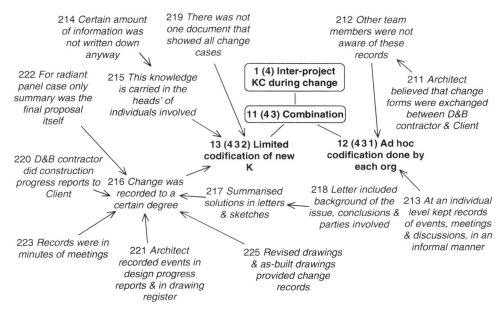

Figure 4.36   Project B: cognitive map of combination stage.

Combination stage

This section addresses the case study findings for the combination stage, along with the relevant cognitive map (Figure 4.36).

In terms of codifying the knowledge that was externalised during discussions, the team members appeared not to prioritise codification, and what was codified was done in an ad hoc manner. Change was recorded to some extent in the minutes of meetings and progress reports. Revised and as-built drawings were viewed by the team as a way of recording changes as well. The architect described 'our change records will generally be the drawing register, as we keep a record in that on each drawing revised.' The D&B contractor viewed the letter that it sent to the client with the sketches as a record of the whiteboard change. Similarly, the M&E consultant viewed his letter with the proposal as a record of the radiant panel case. However, these various forms of change records did not provide details of the change process, rather the information was limited to the final change decision. The architect explained this: 'there wasn't any one document showing that these were the variations. What would have been a range of processes.' The architect further explained the reason for limited codification of change processes when he stated, 'there is a certain amount of information that isn't written down and that is within that person's knowledge.'

The changes were recorded in an ad hoc manner. As the architect described, 'we keep our own record of meetings, discussions and on events in an informal manner.' Although the D&B contractor indicated the change record procedures, there was no evidence that changes were effectively recorded in such forms. Indeed, the architect said he was not aware of any change forms. However, he believed there may be such forms exchanged between the D&B

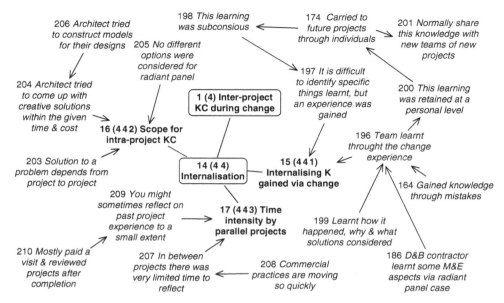

Figure 4.37 Project B: cognitive map of internalisation stage.

contractor and the client. This shows that even if there were change forms in place they were not exchanged effectively between key members.

Internalisation stage

This section addresses the case study findings for the internalisation stage, along with the relevant cognitive map (Figure 4.37).

The team members confirmed that they gained tacit knowledge to a certain extent by going through the considered change cases. For example, the team learnt how and why the whiteboard change occurred. The architect mentioned that he gained knowledge through the mistakes made in the whiteboard case. The team also learnt solutions for both whiteboard and radiant panel cases. The M&E consultant confessed, 'we always learn something new from new projects'. This learning was retained at a personal level and carried to future projects through the individuals involved. As the architect expressed it:

> In the learning process you look at how it arose, why it arose and how you came to solutions. You retain that knowledge on a personal level and use that in future projects.

This learning process, according to the team, was difficult to describe as it was gained subconsciously.

The evidence suggests that the team members (collectively and individually) did not have further opportunities to internalise and reflect upon their experience. They did not consider different options, nor did they undertake experiments to learn further from such problem situations. It was only the architect's role that enabled him to explore creative solutions and to do experiments through design models at the design development stage. Otherwise, due to the time pressures within and between projects, the team members did not have time to experiment and explore what they had learnt from previous projects. The architect stated:

> Commercial practices are moving so quickly. You might sometimes reflect on past project experience, but not to the extent that you are setting here. We generally try, we visit the building, look around and speak to the contractor and the design team and review the complete building and then move to the next one.

Comment on proposition P4 and sub-propositions SP4.1 to SP4.4

> *P4: Project teams are more likely to create new knowledge based on the existing knowledge, during the reactive change process, through a natural flow of knowledge rather than through a full cycle of the knowledge conversion process.*
>
> *SP4.1: Project teams who interact with other team members regularly through face-to-face settings, during the reactive change process, are better at utilising existing tacit knowledge and creating new tacit knowledge, compared to project teams who do not interact regularly through face-to-face settings.*
>
> *SP4.2: Project teams who actively use visualisation techniques during team discussions that arise during the reactive change process, are better at expressing and externalising their tacit knowledge, compared to project teams who do not use such techniques.*
>
> *SP4.3: Project teams are unlikely to effectively combine and codify externalised tacit knowledge arising out of team discussions in reaction change situations.*
>
> *SP4.4: Project teams are more likely to acquire superficial learning through the change experience, during the reactive change process, than effective internalisation through reflection.*

This section summarises the discussions in relation to sub-propositions SP4.1 to SP4.4, and then considers proposition P4.

The findings suggested that tacit-to-tacit knowledge transfer was enabled by socialisation activities between the team members. Thus, sub-proposition SP4.1 is confirmed; namely, project teams who interact with other team members regularly in face-to-face settings, during the reactive change process, are better at utilising existing tacit knowledge and creating new tacit knowledge, compared to project teams who do not interact regularly in face-to-face settings.

According to the case study data, the team showed evidence of externalising their tacit knowledge to explicit knowledge during discussions through different techniques, in particular through the use of diagrams (Table 4.9). Thus, sub-proposition SP4.2 is confirmed; namely, project teams who actively use visualisation techniques during team discussions that arise

Table 4.9   Project B: evidence relating to proposition P4.

Attributes	Evidence	Confirmation/falsification
Presence of socialisation activities	▪ Formal meetings ▪ Specific meetings ▪ Visits to site and offices ▪ Discuss with neighbouring colleagues ▪ Observing other members ▪ Seniors accompanying juniors	Confirmed
Use of different techniques to externalise	▪ Diagrams and sketches ▪ Use of language	Confirmed
Limited combination	▪ Limitation of minutes of meetings, progress reports and drawings as change records ▪ Improper maintenance of change record forms ▪ Failure to prepare project review reports	Confirmed
Superficial internalisation	▪ Learned from mistakes ▪ Learned causes of change ▪ Time intensity between projects ▪ Lack of experiments after change experience	Confirmed

during the reactive change process, are better at expressing and externalising their tacit knowledge, compared to project teams who do not use such techniques.

The case study data suggested that the combination of externalised knowledge was not carried out sufficiently to provide a detailed account of the process as well as the outcome of the change issue. Further, no particular party was assigned to be formally responsible for keeping a record of lessons learnt on change events. This is substantiated by evidence such as lack of details in minutes of meetings, drawings and progress reports; failure to properly maintain change record forms; and failure to prepare a project review report (see Table 4.9). Thus, sub-proposition SP4.3 is confirmed; namely, project teams are unlikely to effectively combine and codify externalised tacit knowledge arising out of team discussions in reaction change situations.

Empirical data further revealed that the team realised learning to some extent through learning from mistakes and causes of change. However, this was not supported by deep internalisation through reflection. This is substantiated by concepts such as time pressures and lack of experiments after the change experience (see Table 4.9). Thus, the case study findings confirm sub-proposition SP4.4; namely, project teams are more likely to acquire superficial learning through the change experience, during the reactive change process, rather than effective internalisation through reflection.

Figure 4.38   Project B: coding structure of inter-project knowledge transfer. Screen shot courtesy of QS International.

On the whole, the sub-propositions SP4.1 to SP4.4 confirm that the team gained new knowledge through interactive settings while going through the change experience. However, a full cycle of knowledge conversion from socialisation to internalisation via externalisation and combination modes was not apparent. The team gained tacit knowledge through socialisation and internalisation, but team members failed to combine the externalised knowledge properly. Thus, proposition P4 is confirmed; namely, project teams are more likely to create new knowledge based on the existing knowledge, during the reactive change process, through a natural flow of knowledge rather than through a full cycle of the knowledge conversion process.

4.3.6   Inter-project knowledge transfer via change

Figure 4.38 provides the coding structure that relates to inter-project knowledge transfer. This is discussed through two parts: first, transfer of knowledge via individuals; and, second, inter-project knowledge dissemination. Finally, a summary of the findings and comments on proposition P5 and sub-propositions SP5.1 to SP5.2 are provided.

Transfer of knowledge via individuals

The following discussion on the case study findings is presented, along with the cognitive map in Figure 4.39.

The case study data revealed that the team developed its tacit knowledge base by going through the 'learning by doing' experience of change cases.

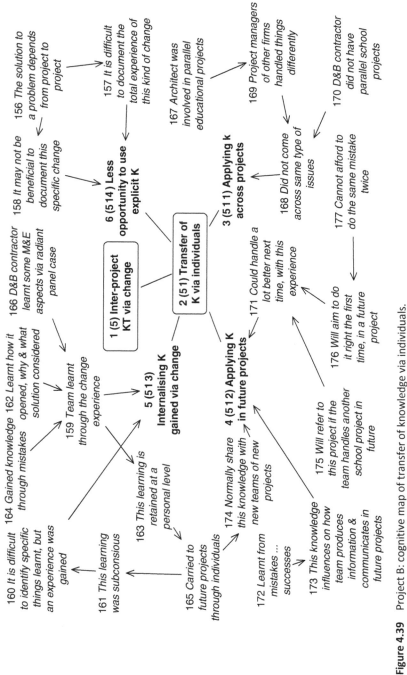

Figure 4.39 Project B: cognitive map of transfer of knowledge via individuals.

This was discussed under the internalisation stage. The knowledge gained through the change experience was mainly held in the heads of the team members and carried forward to future projects. The architect mentioned this on two instances: 'you retain that knowledge [change experience] on a personal level and use that in future projects', and, 'there is a certain amount of information that isn't written down anyway and just contains in the head'. According to the team members, this knowledge will enable them to handle future projects better. The D&B contractor said that in future projects he will make sure not to make the same mistake again. The architect stated that he will make use of his new knowledge if he handles a school project in future.

On the other hand, the knowledge gained through change experience does not seem to have been transferred to future projects through codification strategies. This is explained by the relative absence of opportunities to use explicit knowledge, as discussed above. As team members explained, it was difficult (and prohibitively expensive) to write down the total experience of all the specific and nuanced issues involved in a change in a document. Also, it may not be beneficial if other members do not get the opportunity to refer to this codified knowledge in future projects.

Inter-project knowledge dissemination

The following discussion of the case study data is presented, along with the cognitive map in Figure 4.40.

The case study data showed that the project knowledge dissemination at the multiple organisation level was facilitated by face-to-face settings. Even though the in-house D&B team was site-based they still kept close contact with their superiors and colleagues at the head office. For example, there were monthly meetings at the head office where the team had to report project progress to their directors. In addition, the directors visited the site weekly and met the team on an informal basis. The D&B contractor explained how he shared his project knowledge with his colleagues: 'I now [after the school project] talk to people at school project quite a lot, because I know them.' The M&E consultant described how he reached other colleagues at his office to share and gain knowledge. He mentioned, 'if there is a specific problem I always reach an expert in the company to discuss and get an opinion.' The architect also gave an example:

> I was speaking to my director yesterday on an issue regarding a claim that came out of a project – a technical issue. We shared and pooled that knowledge.

At his office the architect got together with his directors and other project architects and discussed project matters. He further mentioned that in his practice he tries to introduce brainstorming activities for design sessions depending on the pressures within the practice. The team members also participated in social events, where they got an opportunity to meet other

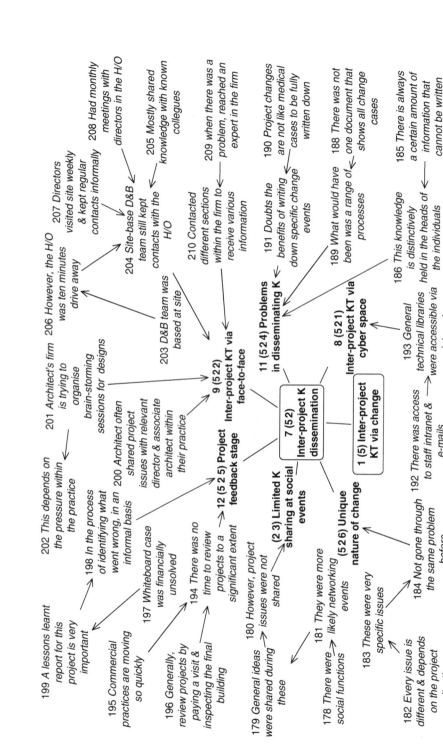

Figure 4.40   Project B: cognitive map of inter-project knowledge dissemination.

colleagues. However, these were viewed by the interviewees more as networking events than specifically to share project experiences.

As explained above, there was no evidence of codifying project experience to disseminate to a wider audience. The D&B contractor identified the importance of a 'lessons learnt' document for the school project. But the project did not get sufficient time to prepare this document as the team moved on to new projects by the completion stage of the school project. The architect, too, mentioned the absence of any document that showed a detailed account of change cases. According to him, there is always a certain amount of information that cannot be written down and that is carried in the heads of the individuals involved. The M&E consultant further described the problems of disseminating such documents. He stated:

> I do not believe that it is beneficial to document this kind of change. It is not a change that could be written in a report, such as a medical change for another person to read. The solutions to the problem change from project to project.

The unique nature of this type of change makes it more difficult to disseminate that knowledge through codification strategies.

The team members had some opportunities to share and disseminate information through cyberspace. They had e-mail facilities and access to an intranet where the common knowledge such as technical libraries was stored. The architect expressed the limitations of sharing knowledge this way:

> … we have technical libraries that would perhaps have the information needed. How efficient are they? – that is really the question. They were really not up to the required level.

More importantly, the specific knowledge and lessons learnt from previous projects are not disseminated through these channels.

Comment on proposition P5 and sub-propositions SP5.1 and SP5.2

> P5: *The new knowledge that is created during the reactive change process is transferred to multiple organisations, for potential re-use in future projects, through personalisation strategies, rather than through codification strategies.*
> SP5.1: *The new knowledge created during the reactive change process is re-used in future projects through the individuals involved during the process, rather than through the codified documents.*
> SP5.2: *The new knowledge created during the reactive change process is disseminated and made available to the wider organisation for potential future re-use, through interactive settings between the organisational members, rather than through effective dissemination of codified documents.*

Table 4.10   Project B: evidence relating to proposition P5.

Attributes	Evidence	Confirmation/ falsification
Transfer to future projects via individuals	▪ Less opportunity to use explicit knowledge ▪ Members applying knowledge in future projects	Confirmed
Dissemination via personalisation	▪ Monthly meetings ▪ Directors' site visits ▪ Social events ▪ Brainstorming sessions ▪ Discuss with colleagues ▪ Absence of specific knowledge in intranet	Confirmed

The knowledge gained through a change experience is more likely to be passed on to a similar context through the individuals involved. The team members tended to apply this knowledge in future projects by handling such issues better and not making the same mistake twice. Further, the data explained the reduced opportunity to use explicit knowledge (Table 4.10). Thus, sub-proposition SP5.1 is confirmed; namely, the new knowledge created during the reactive change process is re-used in future projects through the individuals involved during the process, rather than through the codified documents.

The case study data suggested that at the firm level there were more opportunities to disseminate project knowledge through informal face-to-face settings rather than through formal codification strategies. Thus, sub-proposition SP5.2 is confirmed; namely, the new knowledge created during the reactive change process is disseminated and made available to the wider organisation for potential future re-use, through interactive settings between the organisational members, rather than through effective dissemination of codified documents.

Considering sub-propositions SP5.1 and SP5.2, the case study suggests that new knowledge created during change events passed to future projects through personalisation strategies rather than codification strategies. Thus, proposition P5 is confirmed; namely, the new knowledge that is created during the reactive change process is transferred to multiple organisations, for potential re-use in future projects, through personalisation strategies, rather than through codification strategies.

4.4   Summary and Link

This chapter has presented key findings from the selected case studies to test the proposition developed in Chapter 2, using the research methodology set out in Chapter 3. The case study Project A was in a partnering arrangement. A 'change in floor design', which was identified as a significant change event in this project, was selected for detailed investigation. The case study Project

B was in a D&B arrangement. The change events selected in this project were a 'change in height of whiteboards' and a 'change in design of radiant panels', which had significant impacts on the project. The findings of both case studies, Project A and Project B, confirmed propositions P1 to P5. The next chapter presents the cross-case analysis of the Project A and Project B case studies.

Cross-Case Analysis

5.1 Introduction

Chapter 4 discussed the research findings of change case studies. The aim of this chapter is to present the findings of the cross-case analysis, which identifies similarities and differences. The chapter is structured as follows:

- First, the findings relating to the contextual factors of reactive change are discussed (see Section 5.2).
- Second, the findings relating to the properties of reactive change are discussed (see Section 5.3).
- Third, the findings relating to the identification and utilisation of knowledge during reactive change are discussed (see Section 5.4).
- Fourth, the findings relating to the creation of knowledge through reactive change are discussed (see Section 5.5).
- Fifth, the findings relating to the transfer of knowledge through reactive change are discussed (see Section 5.6).

5.2 Contextual Factors of Change: P1 Discussion

P1: The reactive change process is very complex and ill-structured rather than simple and well-structured, due to various contextual factors that could be identified under process characteristics, group characteristics, organisational characteristics and wider-environmental characteristics.

The case studies identify several factors grouped under process characteristics (see Chapter 4). The data suggest that changes are caused mainly by design-driven triggers. For example, errors and omissions made at the design phase led to significant changes in the construction phase. These findings are consistent with the construction project change literature in Chapter 2.

Love and Li (2000), for example, note that a higher percentage of changes originate from the design phase compared to the construction phase.

From the empirical data the early identification of change emerged as a significant enabler. This is consistent with decision-making theory (see, e.g., Simon, 1957; Mintzberg *et al.*, 1976), which argues that a problem-solving process should start with an adequate problem recognition stage (see Section 2.3.3). Changes identified early minimised disruption. The case studies show how late identification of change adversely affects project tasks and creates a series of disruptive effects. The construction literature frequently confirms this point. For instance, CIRIA (2001) shows that the impact of changes increases when a project moves from design to construction phases (see Section 2.3.2).

Once the need for a change is realised, the team members need to efficiently address the issue. The case studies stress how lengthy decision-making processes had considerable impact on project progress. Especially in case study Project A, the client's slowness in making its change decision significantly and adversely influenced the whole process. This was mainly due to the nature of the client organisation; the in-house decision-making process of a multi-headed client being political and hierarchical; and the representative of the client organisation who participated at team discussions not being given sufficient authority to make prompt decisions. The findings emphasise the importance of the client role in making prompt decisions in relation to unplanned changes (see, e.g., Akinsola *et al.*, 1997).

The case studies demonstrate that project changes were generally made complex by 'social processes' between team members rather than by the technical complexity of the issue itself. This was an emergent finding from the empirical data that is not directly addressed in the prevalent literature.

The findings indicate that the cost significance of project change is considered more influential than the time and quality significance. Generally, project teams attempt to meet originally established quality standards and time targets by employing more people and resources. However, the cost of change has to be negotiated between the relevant parties. The case studies show how disputes can occur due to financial implications and how this affects the whole managing change process. In fact, the way project team members view project change can depend on who bears the cost of change. For example, a change can be beneficial to the client but disruptive to the contractor if the latter has to bear the cost.

In terms of group characteristics (see Chapter 4) the significant factors that affect change situations have been discussed under team culture, team communication, team consistency and team leadership.

The partnering arrangement in case study Project A showed a positive knowledge-sharing culture and good teamworking environment between all key members. With prior relationships developed through working together in previous projects, the team members were trustworthy, dependable and reliable to one another. Partnering arrangements, however, were shown to have certain limitations. For example, findings from case study Project A demonstrated the difficulty of bringing new members into a long-established system. This resonates with Leonard-Barton (1995), who argued that, over time, core capabilities may become core rigidities. Applying this concept, capabilities such as long-term learning curves and good teamworking in collaborative team arrangements can, in themselves, lead to rigidities, such as not allowing entry of new members and thereby new knowledge into the system (see Section 2.5.3).

Conversely, the absence of a partnering arrangement and thereby prior relationships in case study Project B led to ineffective knowledge-sharing. Even though the members were in a D&B arrangement, divisions existed between parties especially between the client team and the D&B team. This led to a 'blaming culture' and inflexibility when it came to negotiations and miscommunication between parties. This finding is consistent with the construction literature (see Section 2.2.3), especially with Moore and Dainty (1999), who state that, in practice, D&B arrangements still exhibit traditional professional divisions.

Further, in terms of team culture, members seem to share a common language. In case study Project A with its partnering arrangement, the client too shared construction knowledge to a considerable extent with the rest of the project team. The extent of using a common language, however, seems to be influenced by the technical complexity of the issue. If the change relates to general construction the whole team seems to understand and pool their knowledge, whereas if the issue is of a specialist nature, such as a refrigeration issue, knowledge sharing is limited between the members who possess the narrow expertise base.

The existence of proper change management processes, which define the role of each participant, is essential to efficiently solve a change problem. The case study data reveal the heavy reliance on regular meetings in communicating and negotiating change issues between team members. This reliance on face-to-face settings (rather than technology-enabled communication such as e-mails and intranets) is consistent with the literature. For example, Luray and Raisinghani (2001) emphasise that despite the important role that the technology plays in connecting virtual teams, technology alone is insufficient and face-to-face interactions are still essential to promote effective teamworking. Indeed, Koskinen et al. (2003) argue that face-to-face interaction is the richest medium because it allows immediate feedback (see Section 2.4.2).

A consistent team during change events was found to be a key variable in successfully managing project change. The change of client representatives at meetings in case study Project A, and the change in the D&B firm's management team in case study Project B, adversely impacted on the change decision-making process. According to case study Project B, when key members left a project in between change issues, a sufficient transfer period between old and new members needed to be considered in order to minimise the disruptive effects. Team migration effects can be further minimised if new members maintain links with the old members. The literature has addressed similar findings (see Section 2.4.2). Chapman (1999), for instance, identifies the highly disruptive influence when a team member departs in the middle of a project. He stressed that the knowledge of the historical development of the project will be lost, and as project information is often

Table 5.1   Contextual factors: enablers and barriers.

Attributes	Enablers	Barriers
Process characteristics	▪ Early identification ▪ Efficient client decision-making process	▪ Omissions and errors ▪ Improper managing of change processes ▪ Multi-headed client ▪ Process complexity ▪ Uncertainty
Group characteristics	▪ Partnering arrangement ▪ Regular meetings ▪ Effective use of IT	▪ Inconsistent team ▪ Blaming culture
Organisation characteristics	▪ Site decisions uninfluenced ▪ Top management support	▪ Absence of rewards ▪ Time intensity

voluminous and complex it cannot be passed totally from one individual to the next, even if there is a handover period.

The influence of organisational structure and culture was discussed under organisational characteristics of project changes (see Sections 4.2.2 and 4.3.2). The partnering arrangement in case study Project A shaped the culture and structure of dealing with project changes. There was minimum influence from the organisation level to project-level decisions. Top management support was given only if required by the project team members. Case study Project B, despite the absence of a partnering arrangement, also had a change management culture. The findings suggest that generally project changes are managed independently at site level and top management only gets involved in exceptional circumstances.

The existence of a learning culture at the organisation level positively influences the change management system. In the case study firms, members were generally given autonomy and freedom to absorb knowledge. However, there were no explicit rewards, especially financial rewards, to promote knowledge creation and learning. On the other hand, time pressures constrained individuals from searching for innovative ideas and considering knowledge production.

In summary, the findings indicate that there are a number of project environment factors that impact on project change. These various factors demonstrate the complex and ill-structured nature of a project change situation. Hence, proposition P1 is confirmed; namely, the reactive change process is very complex and ill-structured rather than simple and well-structured, due to various contextual factors that could be identified under process characteristics, group characteristics and organisational characteristics. However, both case studies failed to capture factors related to wider-environmental characteristics.

Building on these contextual factors that are identified in this cross-case analysis, Table 5.1 summarises the key enablers and barriers that provide lessons for effective project change management.

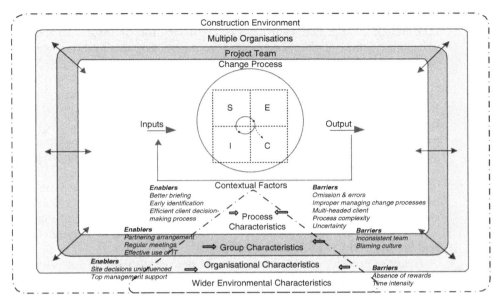

Figure 5.1 Contextual factors: enablers and barriers. S, socialisation; E, externalisation; C, combination; I, internalisation.

Figure 5.1 maps these factors onto the conceptual model that was developed through the literature review (see Section 2.6) to provide a better view of proposition P1.

5.3   Knowledge Properties During Change: P2 Discussion

P2: The knowledge that the project team members use in the reactive change process is: more tacit than explicit; more collective (mutual) than individual; and more situational than prompted.

The construction team in the case study Project A comprised an experienced group of professionals who possessed substantial knowledge of general construction processes. This finding was replicated in case study Project B. The Project A team showed evidence of more experience in the specific project type compared to the Project B team. This was mainly due to the partnering arrangement that was present in Project A. Partnering enabled the same team members to work in the same type of projects for a longer period. However, the absence of a partnering arrangement in Project B benefited that team in a different way. The experience of working with different teams over different project types enabled the second team to gain a wider range of expertise. This finding can be traced back to the literature review, which addressed the issue of exercising collaborative team arrangements cautiously (see Section 2.5.3), in that long-term partnering arrangements can lead to core rigidities from core capabilities (Leonard-Barton, 1995) and constrain the search for novel solutions (Hansen, 1999). On the other hand, other contractual arrangements can provide cross-fertilisation instead of specialisation, by rotating and exposing construction participants to different members through varied projects.

The cases show that long-term experience, either specialised or diverse, significantly influences the ability of project teams to solve change problems. According to both case studies, construction project teams relied heavily on tacit knowledge based on their past experience. In both case study projects, interviewees expressed that they generally had to deal with a new project situation where they had to find a quick practical solution. Team members relate their past experience with the new situation and come up with practical solutions. The solution-generation process based on past experience, according to interviewees in both case study projects, was subconscious and difficult to make transparent. This can be related to the semiconscious and unconscious nature of tacit knowing (Polanyi, 1966; Leonard and Sensiper, 1998) as discussed in Section 2.5.1.

The need to find practical solutions in new project situations restricted the case study project teams from referring to previous knowledge codified in documents. In effect, previous codified lessons learnt are not available to members in an appropriate format for easy retrieval. Team members often do not know where to look in previous project files. This finding is consistent with the general project-based literature (see Section 2.4.2). For example, Disterer (2002, p. 513) states, 'in most cases, even the place where the documentations of a specific project is stored will be unknown'; and, according to Koskinen *et al.* (2003), in practice a team member often relies upon other team members for knowledge and advice, rather than turning to databases and procedure manuals. They could take advantage of codified documents only when colleagues direct them to a specific point. The teams, though, referred to current project documents such as drawings, specifications, minutes of meetings and other correspondence in order to identify the extent of the problem.

Overall, the above discussion suggests that construction project teams rely on tacit knowledge rather than explicit knowledge in managing unplanned change events. This finding is consistent with the general construction literature (see Section 2.4.2). For example, Gann and Salter (2000) state that tacit knowledge is extremely important within the construction environment; and Li and Love (1998) stress that construction problem-solving significantly relies on experiential knowledge.

According to both case study projects (see Sections 4.2 and 4.3), the construction project teams managed change collectively rather than individually, and construction tasks are dependent on each project member's role. This task dependency strengthens the literature findings (see Section 2.2.2) that in construction projects there exists a 'team' rather than a 'group' of professionals with complementary skills that need to be put together towards the project goals (Cornick and Mather, 1999). The team set-up gives everybody a chance to put their views across at team discussions and to reach a consensus decision. This finding is consistent with the extant literature (see Section 2.2.4), which states that construction problem-solving

Table 5.2   Knowledge properties: enablers and barriers.

Attributes	Enablers	Barriers
More tacit, collective, situational	▪ Long-term experience ▪ Team decision-making ▪ Need for a practical solution	▪ Single focused experience ▪ Different individual interests ▪ No motivation for innovation

takes place in a team environment (see, e.g., Gunasekaran and Love, 1998; Anumba *et al.*, 2001). Further, this finding can be related to Leonard and Sensiper's (1998) claim that the purest form of collective tacit knowledge exists within teams, which are characterised by complementary individual knowledge bases and bonds of shared accomplishment (see Section 2.5.1).

The findings from both case study projects (see Sections 4.2.3 and 4.3.3) revealed that in managing unplanned changes within construction projects the primary focus was to quickly and practically solve the problem rather than consider more long-term innovative ideas. Knowledge creation was not the primary aim during construction problem-solving; rather, it was a welcomed 'by-product'. Knowledge used in managing project change is situational rather than prompted. The prevalent literature strengthens this finding (see Section 2.4.2). For example, Sexton and Barrett (2003, p. 615) stress that 'innovation often takes the form of pragmatic problem-solving on site that could not have been reasonably predicted before the project started'.

In summary, the discussion in this section confirms proposition P2; namely, the knowledge that the project team members use in the reactive change process is: more tacit than explicit; more collective (mutual) than individual; and more situational than prompted.

Based on these findings, Table 5.2 provides enablers and barriers that enhance or impede the above identified knowledge properties. These enablers and barriers are mapped onto the conceptual model (Figure 5.2) to give a better picture of proposition P2.

5.4   Knowledge Identification and Utilisation During Change: P3 Discussion

> *P3: Project team members who know where the knowledge resides within the team through prior interactions, are better at identifying and actively utilising relevant team members' knowledge, during the reactive change process, compared to project team members who do not know where the knowledge resides within the team.*

Case study A (see Section 4.2.4) showed that the project team was able to generate several alternative solutions during managing reactive change by effectively identifying and utilising relevant members' knowledge. This was due to the good teamworking environment that had been fostered by the

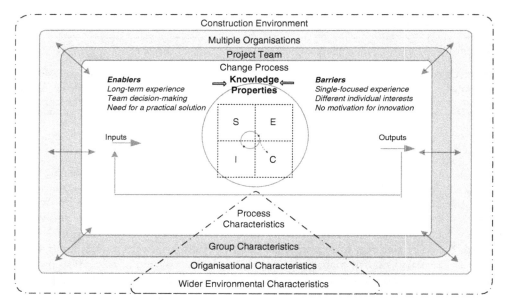

Figure 5.2   Knowledge properties: enablers and barriers. S, socialisation; E, externalisation; C, combination; I, internalisation.

partnering arrangement. The literature on partnering (see Section 2.2.3) identifies long-term team learning and effective knowledge sharing as key benefits from partnering arrangements (see, e.g., Black *et al.*, 2000; Bresnen and Marshall, 2000; Humphreys *et al.*, 2003).

Case study Project B (see Section 4.3.4) provided a negative case to confirm the same point. It supported the claim that construction teams fail to effectively identify and utilise team members' knowledge when they experience poor relationships. Although the D&B procurement path offered collaborative working, adversarial relations between the client and the D&B team persisted. This finding is consistent with research findings of Moore and Dainty (1999), which state that collaborative approaches, such as D&B projects, are still affected by traditional adversarial cultures (see Section 2.2.4). Good relationships can make team members flexible during negotiations. However, if the issue has a large financial bearing, the open team culture and the flexibility can be affected.

In summary, the findings from both case studies confirm proposition P3; namely, project team members who know where the knowledge resides within the team through prior interactions, are better at identifying and actively utilising relevant team members' knowledge, during the reactive change process, compared to project team members who do not know where the knowledge resides within the team.

Key enablers and barriers for effective identification and utilisation of knowledge, arising from the above findings, are given in Table 5.3. These are mapped onto the conceptual model to give the pictorial view of the proposition P3 (Figure 5.3).

Table 5.3   Knowledge identification and utilisation: enablers and barriers.

Attributes	Enablers	Barriers
Effectively identify and utilise relevant knowledge	▪ Good teamwork ▪ Partnering ▪ Separate meetings for specific issues	▪ Adversarial culture ▪ Significant financial implications

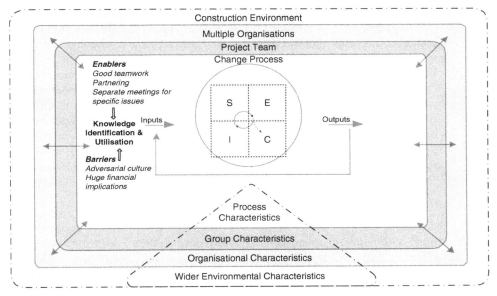

Figure 5.3   Knowledge identification and utilisation: enablers and barriers. S, socialisation; E, externalisation; C, combination; I, internalisation.

5.5   Intra-Project Knowledge Creation During Change: P4 Discussion

> *P4: Project teams are more likely to create new knowledge based on the existing knowledge, during the reactive change process, through a natural flow of knowledge rather than through a full cycle of the knowledge conversion process.*
>
> *SP4.1: Project teams who interact with other team members regularly through face-to-face settings, during the reactive change process, are better at utilising existing tacit knowledge and creating new tacit knowledge, compared to project teams who do not interact regularly through face-to-face settings.*
>
> *SP4.2: Project teams who actively use visualisation techniques during team discussions that arise during the reactive change process, are better*

at expressing and externalising their tacit knowledge compared to project teams who do not use such techniques.

SP4.3: Project teams are unlikely to effectively combine and codify externalised tacit knowledge arising out of team discussions in reaction change situations.

SP4.4: Project teams are more likely to acquire superficial learning through the change experience, during the reactive change process, rather than effective internalisation through reflection.

As discussed in Section 5.3, both case study projects indicated that construction team members shared experiences with each other when managing change events. When managing unplanned changes the teams preferred and held regular face-to-face meetings. Project changes were discussed at formal progress meetings and specific issues relating to these changes were discussed at separate meetings. There occurred informal discussions, as when visiting other members' offices and the project site to discuss issues.

During these face-to-face settings team members had the opportunity to observe and learn from other team members who were experts in different disciplines. This suggests that horizontal learning across disciplines existed during managing change. This finding is consistent with the construction literature that addresses team learning (see Section 2.4.2), in that several researchers have identified that team learning especially takes place during regular meetings involving different participants (Huber, 1996; Barlow and Jashapara, 1998; Busseri and Palmer, 2000).

The cases showed that there was capacity for learning within a particular discipline during change events. For example, when seniors accompanied juniors during problem-solving activities, vertical learning occurred. Construction professionals also learned from their colleagues who were seated close to them in their offices and who happened to work on similar types of project. The site-based project team members in case study Project B also learned from one another by sharing their knowledge relating to project change. Hence, seating arrangements within offices had an influence over how members shared their project experiences. There existed other forms of gaining tacit knowledge during change situations; for example, when members visited similar sites and participated in social events. These forms of tacit knowledge utilisation are identified in the general knowledge management literature (see Section 2.5.2); according to Nonaka *et al.* (1994), the socialisation mode can be evident through the presence of joint activities, apprenticeships, informal networks, social events, face-to-face settings and prior interactions.

The above discussion suggests that during unplanned change situations team members are exposed to socialisation activities that enable them to create new tacit knowledge. Thus, sub-proposition SP4.1 is confirmed; namely, project teams who interact with other team members regularly through face-to-face settings, during the reactive change process, are better at utilising existing tacit knowledge and creating new tacit knowledge, compared to project teams who do not interact regularly through face-to-face settings.

The team members externalise their tacit knowledge in different ways at discussions. According to both case studies the most preferable technique was to visualise thoughts through pictures, diagrams and sketches. This reliance on visualisation techniques to externalise thoughts is consistent with the construction literature. For example, Perry and Sanderson (1998) claim that design is an interactive activity where multiple actors represent their thoughts through artefacts such as plans, models, prototypes and visualisations (see Section 2.4.2). In addition, project team members, where appropriate, used examples of previous projects. During discussions the team members generally listened and encouraged each other to express their views. When their thoughts were not clear they requested them to either be more specific, repeat or explain through a diagram. The discussions were more like brainstorming sessions where teams attempted to generate change options.

These findings confirm sub-proposition SP4.2, namely, that project teams who actively use visualisation techniques during team discussions that arise during the reactive change process, are better at expressing and externalising their tacit knowledge compared to project teams who do not use such techniques.

The use of different techniques is dependent on individual members and their specific roles. For example, architects prefer visualisation of their ideas through drawings, a preference inspired by their functional role. The quantity surveyors prefer to use language to express ideas, whereas the contractors prefer to use examples from previous projects that they have built. Further, how members externalise thoughts depends on the nature of the issues. For instance, if the issue has financial implications the externalisation stage can suffer as each member tends to be careful when they express ideas.

Externalised ideas and thoughts during discussions do not seem to be effectively codified during the managing change process. Interviewees of both case study projects (see Sections 4.2.6 and 4.3.6) viewed minutes of meetings as a record of changes. However, rather than recording the total change process, the information contained in the minutes of meetings was limited to agreed change decisions – not the underpinning rationale on why and how these decisions were arrived at. Construction project teams also exchange change record forms, which provide instructions with details of the finally agreed change option. These forms were not always properly maintained. Change records also took the form of revised drawings or as-built drawings. These, again, provided details of the final change option rather than a detailed account of the change process. Sometimes records of project changes were found in letter correspondence and progress reports in an ad hoc form. Furthermore, neither of the case study projects showed evidence of a project review report that pulled together the lessons learnt. The limited codification and lack of details in the project documents is consistent with other project-based literature (see Section 2.4.2). For example, Schindler and Eppler (2003) state that the relevant project documentation is often superficial and lacks records such as reasons for failure, and how solutions were built and implemented. Disterer (2002, p. 516) expresses similar views: 'the documentation of projects rarely contains knowledge for following projects'.

Further, no particular party was made responsible to maintain the overall lessons learnt of the project. What generally happened was that individual members kept their own records of the process in an ad hoc manner. This raises the ownership issues as discussed in Section 2.4.2, in that Chan *et al.* (2004) argue that in a construction project setting the leadership and knowledge ownership are unclear, which creates problems as to which party is responsible for capturing and storing knowledge.

Both case study projects confirm sub-proposition SP4.3, namely that project teams are unlikely to effectively combine and codify externalised tacit knowledge arising out of team discussions in reaction change situations.

In terms of internalising knowledge created during change, both projects revealed that team members increased their tacit knowledge-base as they went through the change experience. Project teams generally learnt technical issues, causes of change, solutions for change and, more importantly, how to deal with people during the process. The team members built upon their mistakes and developed a better understanding of change situations. This enabled them to effectively deal with future changes and, more importantly, to go through the process.

However, the data suggested that this internalisation process was limited to immediate learning and was not further strengthened through reflection and experimentation. Project team members moved to the next project immediately after a project and sometimes they worked in parallel projects, which gave them very little time to reflect. This finding is consistent with the general construction literature (see Section 2.5.3). For example, Disterer (2002) stresses that the project review stage often has to be dropped because of time constraints and members immediately moving on to other projects.

The above discussion shows that in managing change situations construction projects gain tacit knowledge through the experience. But further reflections from these lessons learnt are constrained by time pressures. This confirms sub-proposition SP4.4, namely that project teams are more likely to acquire superficial learning through the change experience, during the reactive change process, rather than effective internalisation through reflection.

In summary, the sub-propositions confirmed above suggest that a full knowledge conversion cycle, as Nonaka and Takuechi (1995) explain (see Section 2.5.2), does not arise during managing unplanned change within construction projects. Even though there is evidence of new knowledge creation during the process, it is difficult to identify the conversion of tacit knowledge into explicit knowledge. What generally happens is that team members exchange their tacit knowledge gained from previous experience by socialising in face-to-face settings. The tacit knowledge that resides in individuals' heads could become explicit during these discussions and may sometimes be recorded in project documentation. Team members internalise this whole experience and increase their tacit knowledge-base, which they carry to future projects. The findings reveal that tacit knowledge is playing a key role during the whole process. This knowledge creation process is consistent with Snowden's (2002) explanation of natural knowledge flows, where knowledge can be created without necessarily going through a codification stage (see Section 2.5.2). Thus, the findings confirm proposition P4, namely that project teams are

Table 5.4 Intra-project knowledge creation: enablers and barriers.

Attributes	Enablers	Barriers
Socialisation	■ Progress meetings ■ Specific meetings ■ Visits to sites and offices ■ Accompanying juniors ■ Seating arrangements	■ Lack of face-to-face settings ■ Key members distanced
Externalisation	■ Visualising via diagrams ■ Use examples ■ Listen and encourage ■ Brainstorming sessions	■ Disputes between parties ■ Sticky nature of knowledge ■ Individual characteristics
Combination	■ Record details in minutes of meetings, change order forms, letters, drawings and project review report ■ Responsible party to codify	■ Lack of details in records ■ Ad hoc recording of lessons learnt
Internalisation	■ Time to experiment ■ Client requesting to consider innovative ideas ■ Build on mistakes and failures	■ Time pressures ■ Lack of motivation for innovation

more likely to create new knowledge based on the existing knowledge, during the reactive change process, through a natural flow of knowledge rather than through a full cycle of the knowledge conversion process.

The findings, as discussed under this proposition P4, suggest key enablers and barriers that affect socialisation, externalisation, combination and internalisation stages of a knowledge conversion cycle. These are summarised in Table 5.4.

As Figure 5.4 depicts, these enablers and barriers are subsequently mapped onto the conceptual model to offer a better picture and locate proposition P4 within the research problem.

5.6   Inter-Project Knowledge Transfer During Change: P5 Discussion

P5: The new knowledge that is created during the reactive change process is transferred to multiple organisations, for potential re-use in future projects, through personalisation strategies rather than through codification strategies.

SP5.1: The new knowledge created during the reactive change process is re-used in future projects through the individuals involved during the process, rather than through the codified documents.

SP5.2: The new knowledge created during the reactive change process is disseminated and made available to the wider organisation for potential future re-use, through interactive settings between the organisational members, rather than through effective dissemination of codified documents.

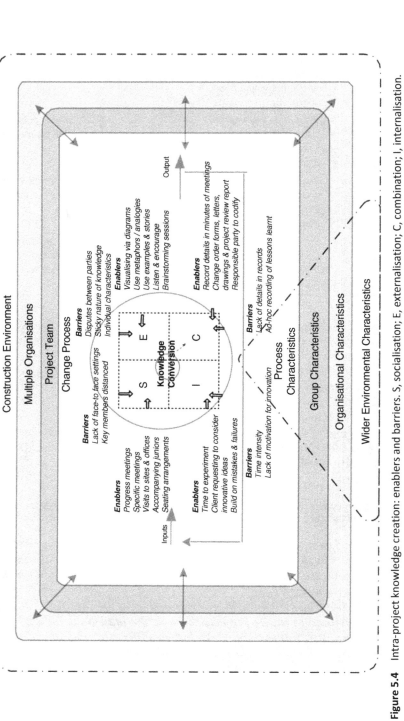

Figure 5.4 Intra-project knowledge creation: enablers and barriers. S, socialisation; E, externalisation; C, combination; I, internalisation.

The case studies show that the knowledge that construction project teams learn from change experience passes to other projects through the individuals involved, rather than through codified project documents. As discussed in the previous section, there was evidence that the project team members learnt from change experiences. This learning is held within the heads of the project participants as tacit knowledge and is passed to parallel and future projects through them. Application of knowledge in future projects, however, is found to be influenced by other variables. Sometimes specific changes may not be experienced twice by the same individual. This can be further influenced by the extent that construction project team members have opportunities to work in similar types of project and experience similar issues.

On the other hand, there was less opportunity for team members to refer to codified documents of past projects. One reason was limited availability of codified documents, and another was limited details in the available codified documents. Therefore, team members were reluctant to use explicit knowledge that was generated through change events and relied heavily on their personal knowledge. This confirms sub-proposition SP5.1, namely that the new knowledge created during the reactive change process is reused in future projects through the individuals involved during the process, rather than through the codified documents. As discussed in Section 2.4.2, this view is strengthened by the construction literature, which states that knowledge generated within projects is mostly limited to individuals (Barlow and Jashapara, 1998; Winch, 2002) and lessons learnt generally become individual tacit knowledge (Gieskes and Broeke, 2000).

Similarly, knowledge that is created during change events passes to other individuals in the multiple organisations through personalisation rather than codification strategies. Both case study projects showed that new knowledge created during project changes were disseminated to a wider community through face-to-face settings originated in the organisation level. The individuals in construction firms met regularly, generally once a month, where they reported their project status before other individuals in the organisation. In addition, there were other forms of face-to-face interaction, which enabled project knowledge to be disseminated to a wider community. For example, annual conferences and social functions, which all provided occasions for knowledge dissemination through personalisation.

Use of codification strategies was limited for the wider dissemination of knowledge. The cases showed that individuals in the organisations had opportunities to share knowledge through cyberspace, but that the knowledge tended to focus on generic issues rather than project-specific ones. This knowledge diffusion problem is addressed in the general construction literature (see Section 2.4.2). For example, Gieskes and Broeke (2000) state that there are no signs of a systematic diffusion of project knowledge within the organisation or across organisational boundaries; and CII (1994) reveals that project databases developed throughout the projects are not effectively maintained at the corporate level for wider dissemination.

Table 5.5   Inter-project knowledge transfer: enablers and barriers.

Attributes	Enablers	Barriers
Transfer to other projects	▪ Face-to-face settings within firms such as meetings, social events, conferences, talking to colleagues and brainstorming events	▪ Specificity of change ▪ Ineffective use of IT ▪ Difficult access to knowledge ▪ Cost of dissemination

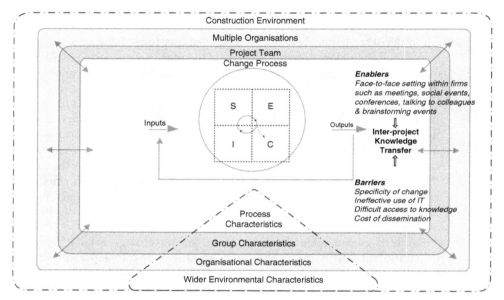

Figure 5.5   Inter-project knowledge transfer: enablers and barriers. S, socialisation; E, externalisation; C, combination; I, internalisation.

The case studies noted that knowledge dissemination through codification strategies was problematic due to several reasons. One was that it can only be beneficial if other members have the opportunity to use this knowledge in future projects. Therefore, when codifying and disseminating very specific changes, organisations have to weigh the benefits against the costs. This is identified in the general knowledge management literature (see Section 2.5.3). For example, Sampler (1998) explains that usability of knowledge transfer depends on how specific and time relevant that knowledge is. Another dissemination problem was the sticky nature of knowledge. A part of knowledge always remains tacit and cannot be codified. The knowledge management literature repeatedly stresses this (see Section 2.5.3). Furthermore, it was difficult to make sure that everybody in an organisation knew what everybody else was doing and where project files were.

The above discussion confirms sub-proposition SP5.2, namely that the new knowledge created during the reactive change process is disseminated and made available to the wider organisation for potential future re-use, through interactive settings between the organisational members, rather than through effective dissemination of codified documents.

In summary, knowledge that is created during change events passes to the organisational level and thereby to future projects through personalisation of individual members rather than through codification and dissemination through information technology. This confirms proposition P5, namely that the new knowledge that is created during the reactive change process is transferred to multiple organisations, for potential re-use in future projects, through personalisation strategies, rather than through codification strategies. This finding is consistent with the general construction literature, especially with the findings of Bresnen *et al.* (2003) (see Section 2.4.2). Bresnen *et al.* (2003, p. 165) state that:

> processes of knowledge capture, transfer and learning in project settings rely very heavily upon social patterns, practices and processes in ways which emphasise the value and importance of adopting a community-based approach to managing knowledge.

Koskinen *et al.* (2003) further strengthen this finding by arguing that, in practice, a team member often relies upon other team members for knowledge and advice, rather than turning to databases and procedure manuals.

Based on the discussion of proposition P5, various factors that enable or hinder effective transfer and dissemination of knowledge from the project change context to the organisational context, are given in Table 5.5. To further explain and understand proposition P5, these factors that enable or hinder inter-project knowledge transfer are mapped onto the conceptual model as illustrated in Figure 5.5.

5.7   Summary and Link

This chapter has presented the cross-case analysis comparing case study Project A with case study Project B. The results indicated that different forms of knowledge were captured, shared and created during problem-solving activities connected with change events when team members interact and discuss, especially through face-to-face settings. This new knowledge created through the reactive change process was generally transferred to multiple organisations through team members and their social networks. The findings were further strengthened by comparing them with extant literature in the broader context. Building upon these findings the analysis further offered enablers and barriers for key variables that related to each proposition. The next, and final, chapter summarises this research, and draws implications from the study for both theory and practice.

Conclusions

Chapter 4 and Chapter 5 presented and analysed the case study findings within the context of the propositions set out in Chapter 2. The aim of this chapter is to summarise the principal findings and provide conclusions and recommendations. The chapter is structured as follows:

- First, a summary of the findings pertaining to each proposition is given (see Section 6.2).
- Second, a comment on the conceptual model is offered with a modified model that represents the findings of this research (see Section 6.3).
- Third, the conclusions about the overall research problem are given (see Section 6.4).
- Fourth, the implications for theory are discussed with the articulation of a knowledge-based project change management model (see Section 6.5).
- Finally, the implications for practice are described with a set of guidelines (see Section 6.6).

This book has investigated the role of team knowledge in 'managing unplanned change situations in the construction phase within collaborative team settings' (referred to in this research as the 'reactive change process'). This overall aim was approached through five research questions (see Section 2.6). First, the research explored key contextual factors of the reactive change process. Second, the properties of knowledge that the project team members use in the reactive change process were investigated. Third, the study focused on exploring how a project team identifies and utilises this knowledge during the reactive change process. Fourth, the study investigated

---

Managing Change in Construction Projects: A Knowledge-Based Approach, First Edition.
Sepani Senaratne and Martin Sexton.
© 2011 by Sepani Senaratne and Martin Sexton. Published 2011 by Blackwell Publishing Ltd.

how project team interaction in reactive change processes generates new knowledge. Finally, the research focused on how any knowledge that is created through the reactive change process is transferred and disseminated within and across multiple organisations for potential re-use in future projects. The research formulated a series of propositions to address the above issues (see Section 2.6). The primary data to test these propositions were collected from three change events that occurred in two construction projects (see Section 3.2.2 in Chapter 3). The case study findings were analysed and presented in Chapter 4 and Chapter 5. The following sections summarise the key findings relating to each proposition.

6.2.1   Summary of proposition P1

> P1: The reactive change process is very complex and ill-structured rather than simple and well-structured, due to various contextual factors that could be identified under process characteristics, group characteristics, organisational characteristics and wider-environmental characteristics.

Various contextual factors of change that related to the reactive change process were identified by this research study in terms of process, group and organisational characteristics.

First, the case studies identified several factors that could be grouped under *process* characteristics. Accordingly, the evidence suggested that reactive changes were generally caused by design-driven triggers (such as errors and omissions made at the design phase). The case studies revealed that late identification of change adversely affected the projects' tasks and created a series of disruptive effects. The general effects of project change were uncertainty for subsequent tasks, and rework, such as design revisions, reprogramming and waste of management time. The case studies further showed that lengthy decision-making processes have considerable adverse impacts on project progress. Thus, the existence of proper change management processes was shown to be essential to efficiently solve change problems. The project changes were generally characterised by 'social complexity' brought about by the individual and collective behaviour, rather than 'technical complexity'. The findings further showed that the cost significance of project change is more influential than time and quality impacts, especially because cost impacts are more likely to create disputes between parties.

Second, the factors relating to *group* characteristics were discussed. The partnering arrangement exhibited in the case study Project A led to a positive knowledge-sharing culture and good teamworking environment. With the prior relationships built through working on previous projects the team members had trust in each other. The partnering arrangement, however, limited the socialisation of new members in the long-standing relationship 'network'. Conversely, the design-and-build (D&B) arrangement in practice showed divisions between parties, especially between the client team and the D&B team, resulting in a 'blaming culture', inflexible parties and poor communication. Team members used a common language of construction technology. Case study data revealed the heavy reliance on regular meetings despite other communication channels. Other important factors were maintaining a consistent team, and providing a sufficient transfer period when the members leave on-going projects.

Third, under *organisational* characteristics, the findings suggested that generally project changes were allowed to be managed independently at site level and top management got involved if required. At the firm level members were not sufficiently motivated to promote knowledge creation and learning. Even though they were given autonomy and freedom to absorb knowledge, explicit rewards (especially financial rewards) were not in place. On the other hand, time pressures eroded the motivation for individuals to search for innovative ideas and generate new knowledge.

To sum up the discussion on proposition P1, the various factors as briefly discussed above (see Section 5.2 for details) explain the complex and ill-structured nature of the reactive change process. Therefore, the proposition P1 was confirmed. The case study sample, however, could not capture wider-environmental characteristics, which demonstrates their rare occurrence and lesser impact on project change.

6.2.2   Summary of proposition P2

> *P2: The knowledge that the project team members use in the reactive change process is: more tacit than explicit; more collective (mutual) than individual; and more situational than prompted.*

The research addressed three properties of knowledge during project change situations. First, the study investigated whether the knowledge a team use in a change situation is more tacit or explicit. Accordingly, specialised experience provided by the partnering arrangement, or the diverse experience provided by other arrangements, was found to significantly influence the solutions of change problems by project teams. The results indicated that construction project teams showed a heavy reliance on their tacit knowledge gained through past experience. Teams generally related their past experience to the new conditions to produce practical solutions. This need to find quick, practical solutions in new project situations constrained project teams from referring to explicit knowledge that is ineffectively codified in documents.

Second, the study looked into the *collective* nature of knowledge during project change situations. Team decision-making was inevitable in a construction project setting, as project members' roles were interdependent. Generally each member was given a chance to put their views across and a consensus decision was reached during team discussions.

Third, the *situational* nature of knowledge was explored. The empirical findings revealed that in managing unplanned changes within construction projects, knowledge creation was not the primary aim, but rather was a by-product.

Taking all these facts into consideration (see Section 5.3 for details), proposition P2 was confirmed.

6.2.3   Summary of proposition P3

> P3: *Project team members who know where the knowledge resides within the team through prior interactions, are better at identifying and actively utilising relevant team members' knowledge, during the reactive change process, compared to project team members who do not know where the knowledge resides within the team.*

The results supported the proposition that a good teamworking environment and prior knowledge of team members facilitated by partnering enabled effective identification and utilisation of relevant members' knowledge during unplanned project change. This helped them to generate several options for the change problem. However, the D&B arrangement alone did not support this claim, due to divisions that existed in parties like in a conventional arrangement.

In summary, the findings (see Section 5.4 for a detailed discussion) confirmed proposition P3.

6.2.4   Summary of proposition P4

> P4: *Project teams are more likely to create new knowledge based on the existing knowledge, during the reactive change process, through a natural flow of knowledge rather than through a full cycle of the knowledge conversion process.*

The primary investigation of this study looked, in turn, into the knowledge conversion stages – namely, socialisation, externalisation, combination and internalisation – to test proposition P4.

In terms of *socialisation*, the case studies indicated that during change events construction team members utilised tacit knowledge by learning from individuals of other disciplines, as well as from seniors within their own disciplines. Teams were exposed to various face-to-face settings (especially meetings) and, in addition, they engaged in activities such as visiting sites, speaking to close colleagues and attending various social events.

With respect to *externalisation*, construction project team members externalised their tacit knowledge in different ways at discussions, especially by visualisation through pictures, diagrams or sketches. Other techniques included the use of examples of previous projects where appropriate. Further, externalisation was supported through the creation of brainstorming environments while listening and encouraging each other.

In looking into the *combination* aspect, the case study results revealed that ideas and thoughts that were expressed at discussions were not effectively codified during the management of the reactive change process. The limited records of project changes (such as minutes of meetings, change order forms, drawings, letter correspondence and progress reports) did not provide a detailed account of change processes. Lessons learnt from project changes were not formally compiled into a project review report and the evidence showed ownership issues of maintaining such documents.

At the *internalisation* stage, evidence supported the claim that the knowledge created during change is internalised by team members. For example, project teams generally learnt technical issues, causes and solutions for change, and dealing with people; and were confident in handling such situations more effectively in future projects. However, as the primary data showed, the internalisation process is not strengthened further through reflection and experimentation, due especially to time pressures.

In summary, the findings (see Section 5.5 for a detailed discussion) confirmed proposition P4.

6.2.5   Summary of proposition P5

> *P5: The new knowledge that is created during the reactive change process is transferred to multiple organisations, for potential re-use in future projects, through personalisation strategies rather than through codification strategies.*

The empirical data confirmed that the knowledge that construction project teams learn from the change experience passes to other projects when the team members engage in parallel and future projects. The tendency is to use the knowledge held within team members' heads in future change events rather than refer to codified documents. The data showed how the limited availability of codified documents and limited details in the available codified documents erode their interest in using the encoded, explicit knowledge in future projects.

Similarly, the findings revealed that new knowledge created during project changes is informally disseminated to a wider community in the multiple organisations through face-to-face settings such as regular meetings, annual conferences, social functions and conversations with colleagues. Use of codification strategies is limited for this wider dissemination of knowledge, as individuals usually shared general knowledge through intranet facilities, rather than specific lessons learnt through project changes.

In summary, the findings (see Section 5.6 for a detailed discussion) confirmed proposition P5.

Overall, the findings offered a series of enablers and barriers in relation to each proposition, as shown in Table 6.1.

The next section discusses conclusions relating to the conceptual model, wherein the above identified enablers and barriers are mapped onto the conceptual model.

6.3   Comment on the Conceptual Model

This section first briefly explains how the conceptual model was developed based on the literature synthesis. Second, it tests and provides the modified model based on the empirical findings that were discussed under each proposition (see Section 6.2).

Table 6.1 Summary of enablers and barriers.

P ref.	Attributes	Enablers	Barriers
P1	Process characteristics	▪ Early identification ▪ Efficient client decision-making process	▪ Omissions and errors ▪ Improper managing change processes ▪ Multi-headed client ▪ Process complexity ▪ Uncertainty
	Group characteristics	▪ Partnering arrangement ▪ Regular meetings ▪ Effective use of IT	▪ Inconsistent team ▪ Blaming culture
	Organisation characteristics	▪ Site decisions uninfluenced ▪ Top management support	▪ Absence of rewards ▪ Time intensity
P2	More tacit, collective, situational	▪ Long-term experience ▪ Team decision-making ▪ Need for a practical solution	▪ Single focused experience ▪ Different individual interests ▪ No motivation for innovation
P3	Effectively identify and utilise relevant knowledge	▪ Good teamwork ▪ Partnering ▪ Separate meetings for specific issues	▪ Adversarial culture ▪ Significant financial implications
P4	Socialisation	▪ Progress meetings ▪ Specific meetings ▪ Visits to sites and offices ▪ Accompanying juniors ▪ Seating arrangements	▪ Lack of face-to-face settings ▪ Key members distanced
	Externalisation	▪ Visualising via diagrams ▪ Use examples ▪ Listen and encourage ▪ Brainstorming sessions	▪ Disputes between parties ▪ Sticky nature of knowledge ▪ Individual characteristics
	Combination	▪ Record details in minutes of meetings, change order forms, letters, drawings and project review report ▪ Responsible party to codify	▪ Lack of details in records ▪ Ad hoc recording of lessons learnt
	Internalisation	▪ Time to experiment ▪ Client requesting to consider innovative ideas ▪ Build on mistakes and failures	▪ Time intensity ▪ Lack of motivation for innovation
P5	Transfer to other projects	▪ Face-to-face settings within firms such as meetings, social events, conferences, talking to colleagues and brainstorming events	▪ Specificity of change ▪ Ineffective use of IT ▪ Difficult access to knowledge ▪ Cost of dissemination

As discussed in Section 2.6, the conceptual model was developed to represent the role of team knowledge during the reactive change process in construction team settings. The change process was considered as an input-transformation-output model. At the input stage, knowledge identification and utilisation for the change process was considered. At the transformation stage the knowledge creation was addressed. At the output stage the transfer and dissemination of knowledge that is created during the change process was taken into account. The predicted knowledge creation cycle was illustrated at the centre. The project-to-project knowledge transfer was represented by arrows that link a project to the multiple organisation layer through the project team layer (see Figure 2.9). The context was shown in four layers: change process; project team; multiple organisations; and construction environment. The characteristics corresponding to each layer (i.e. process characteristics, group characteristics, organisational characteristics and wider-environmental characteristics) were shown along each respective layer.

The appropriateness of this conceptual model in representing the role of knowledge during the reactive change process was strengthened by the empirical investigation. Figure 6.1 shows the modified model by mapping the key findings relating to each proposition onto this model. Accordingly, at the triangle that cuts across contextual layers, key findings in relation to contextual factors (proposition P1) are depicted. Similarly, the other propositions are depicted as follows:

- at the input stage, key findings relating to knowledge identification and utilisation (proposition P3);
- at the transformation stage, the properties of knowledge (proposition P2) and intra-project knowledge creation (proposition P4);
- at the output stage, the inter-project knowledge transfer (proposition P5).

This model is further broadened in Figure 6.2 to represent the various enablers and barriers that were given in Table 6.1.

6.4   Conclusions about the Overall Research Problem

Overall, the aim of this research was to investigate how knowledge is captured, created and used during unplanned change situations in the construction phase within collaborative team settings. The results indicated that different forms of knowledge are captured and shared between project team members during problem-solving activities connected with change events. Generally, team members exchange their tacit knowledge gained from previous experience by socialising through face-to-face settings. This tacit knowledge is often visualised during discussions, when team members share their previous experiences with each other. This visualised tacit knowledge is codified in project documentation in an ad hoc manner. Codification strategies, for example, through information technology, produce only a limited availability of codified documents, with limited detail; this then hinders the transfer and dissemination of new knowledge that is generated through a reactive process. On the other hand, team members internalise

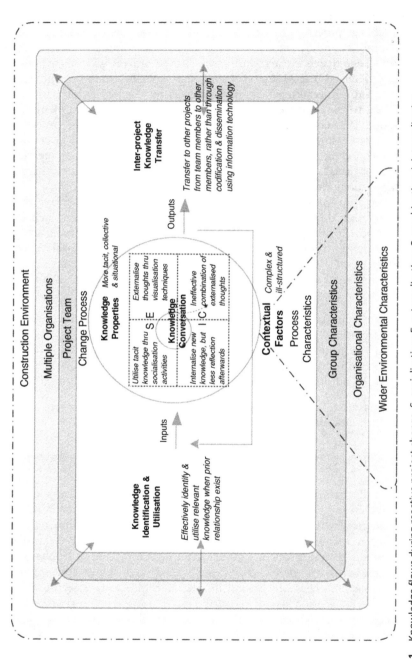

Figure 6.1 Knowledge flows during reactive project change. S, socialisation; E, externalisation; C, combination; I, internalisation.

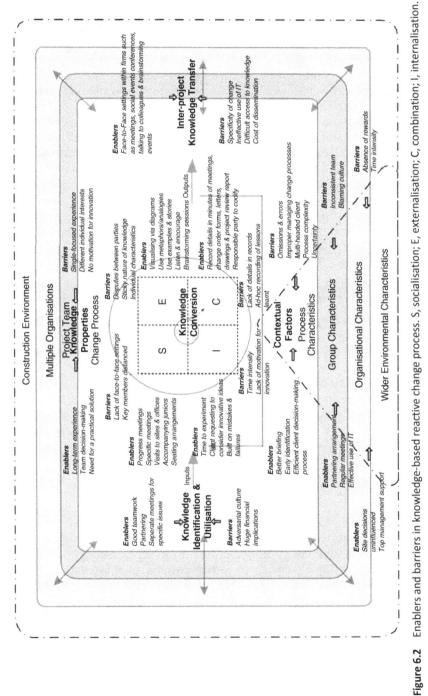

Figure 6.2 Enablers and barriers in knowledge-based reactive change process. S, socialisation; E, externalisation; C, combination; I, internalisation.

new knowledge that is generated through the reactive change process and increase their tacit knowledge base. This new tacit knowledge is generally transferred and disseminated at an organisational level and, thereby, to future projects through personalisation of individual members.

Hence, the overall findings indicate that these knowledge flows are very much centred on tacit knowledge and experience of project personnel. This social construction and use of knowledge in change management challenges the prevailing codification knowledge management solutions based on 'hard' IT approaches, which do not appreciate and accommodate this social phenomenon. Thus, it is argued in this study that there is a need to balance codification knowledge management strategies with 'soft' personalisation strategies to stimulate and support appropriate social interaction between team members and, thereby, enhance the creation, dissemination and shared understanding of tacit project experience. It is through the balance of 'appropriate codification' and 'enhanced personalisation' strategies that collaborative teams can successfully resolve and learn from change events in the construction phase of projects.

With these key conclusions, the next section highlights the implications for theory.

6.5   Implications for Theory

This research contributes to theory by offering a better understanding and a new perspective on the reactive change process. By identifying different variables and thereby introducing enablers and barriers for effective project change management, this research study offers a better understanding of the reactive change process. By integrating theories of knowledge management and project change management, a new perspective on construction project change management is introduced. The next section articulates the specific theoretical contributions by developing a theory of knowledge-based reactive change process.

6.5.1   Theory of knowledge-based reactive change process

According to Nonaka and Takuechi (1995), key to knowledge creation is converting individual tacit knowledge into explicit knowledge. Therefore, they argued that knowledge creation takes place through a knowledge conversion cycle. That is, it proceeds through four modes of knowledge conversion: socialisation, externalisation, combination and internalisation (Figure 6.3). Snowden (2002) criticised this asset-based view of knowledge creation and put forward the claim that knowledge can be created naturally within informal settings, without necessarily converting to explicit knowledge. He described four stages in his 'natural flow of knowledge': complex domain, knowable domain, known domain and chaos domain. In Chapter 2 (see Section 2.5.2), these four stages were approximated to Nonaka's (1994) conceptualisation of the four knowledge conversion modes, while adopting

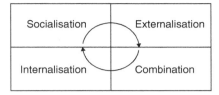

Figure 6.3   Asset-based view of knowledge creation.

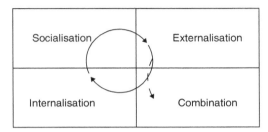

Figure 6.4   Process-based view of knowledge creation.

Snowden's (2002) view of knowledge creation, to identify the process-based view of knowledge creation (Figure 6.4). Thus, Figure 6.3 and Figure 6.4 explain these asset- and process-based views of prevalent knowledge creation theories.

The findings of this study indicated that a full knowledge conversion cycle does not arise during the reactive change process and the knowledge flows are mainly centred on team members' tacit knowledge. Hence, in the context of a reactive change process, knowledge is created and transferred mainly through personalisation of team members, indicating a process-based view of knowledge creation. Therefore, the reactive change process, in general, is consistent with Snowden's (2002) explanation of natural knowledge flows, where knowledge can be created without necessarily going through a codification stage. However, Figure 6.4 does not fully represent the knowledge flows during a reactive change process.

Figure 6.4 does not represent the situational nature of knowledge creation through reactive changes; that is, it does not show that knowledge creation occurs when a project team undergo social interactions to solve a change event that is triggered by a technical issue. The figure does not capture project-based settings and does not show how new knowledge created during a reactive change is fed forward to future projects through multiple organisational settings. Even though the conceptual model developed for this research (see Section 2.6) captured these limitations to a certain extent by representing the reactive change process as a problem-solving process, which occurs in a construction project context, it still denotes Snowden's

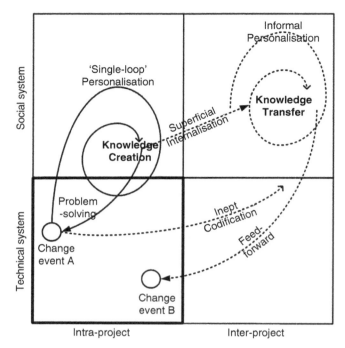

Figure 6.5 Prevalent reactive change process.

(2002) conceptualisation of knowledge creation process at its transformation stage. Taking the closer view of the reactive change process gained from this research, a new theoretical model can be developed, as depicted in Figure 6.5, to better represent the prevalent knowledge flows during the reactive change process.

In Figure 6.5, the technical system denotes the practical, procedural and technological aspects in a construction setting. The social system denotes the networks of construction project participants. To represent the project-based nature of construction, two levels are considered in this model: intra-project and inter-project. Intra-project indicates the activities within a construction project and inter-project indicates activities at the multiple organisation level. The terms personalisation and codification are drawn from Hansen *et al.* (1999) (see Section 2.4.1), to explain knowledge dissemination via person-to-person and person-to-document respectively. Two levels of personalisation are described in the model, by adopting the 'single-loop' and 'double-loop' concepts of Argyris and Schon (1978) (see Section 2.6). Nonaka's (1994) internalisation term is used to describe the experiential learning process (see Section 2.5.2).

The research findings concluded (see Section 6.2.1) that a reactive change is often triggered by a technical issue. Hence, *change event A* (which represents a reactive change) is located within the *technical system* in the *intra-project level*. The research results indicated (see Section 6.2.4) that knowledge is shared and created when project team members interact, to solve the change problem. Thus, this *knowledge creation* process takes place in the *intra-project social system*. The term *'single-loop' personalisation* is introduced here to explain the knowledge dissemination that occurs immediately between project team members when they *solve* the change problem in the *technical system*. Most of the knowledge created remains within the heads of individual project participants. This internalisation process is described in the model as *'superficial'*. This is due to the fact that the new knowledge created through the change experience is simply absorbed by the team members and not subjected to deep reflection and experimentation afterwards.

The results indicated (see Section 6.2.5) that these project team members transfer this internalised knowledge to other individuals when they make informal interactions, at the multiple organisation level. This *knowledge transfer* process is, therefore, located within the *social system* in the *inter-project* level and referred to as *'informal personalisation'*. As the research findings concluded (see Section 6.2.4) the project documentations that had codified the change event generally included details of the final change decisions, but not the details of the whole change experience. Therefore, this codification is depicted in the model as arising from the change event and passing to the *technical system* in the *inter-project level*. This codification does not properly reach the knowledge transfer that happens in the *social system*. Thus, this is referred to as *'inept codification'* in the model. Through the informal personalisation strategies that occur in the *social system* at the *inter-project* level, knowledge is *fed-forward* to future change events (*Change event B* in the model). The model shows that more emphasis is currently placed on the *intra-project technical system*.

Three limitations in the prevalent reactive process are identified to prescribe a knowledge-based approach to effective project change management. First, the 'superficial internalisation' needs to be followed by deep reflection and experimentation. Second, 'informal personalisation' at the inter-project level needs to be recognised and strengthened to create an effective personalisation for wider knowledge dissemination. Third, an appropriate codification that codifies newly created knowledge (such as solutions, lessons learnt) and transmits it to the social system at the inter-project level, should be adopted. Following these prescriptions, Figure 6.6 represents the 'knowledge-based reactive change process'.

Figure 6.6 introduces 'deep internalisation' as opposed to 'superficial internalisation'; 'double-loop personalisation' compared to 'informal personalisation'; and 'apt codification' instead of 'inept codification' for effective project change management. For *deep internalisation*, team members need to reflect deeply on their change experience and carry out further experiments where possible. For this to happen, team members need to be given sufficient time and resources, by project clients and top management of individual organisations. To enhance personalisation, the management at multiple organisation levels needs to recognise and strengthen social networks and face-to-face settings. To achieve *apt codification*, knowledge created during the reactive change process need to be codified and transmitted to the *social system* where knowledge dissemination takes place through personalisation. The 'deep internalisation', 'apt codification' and strengthened

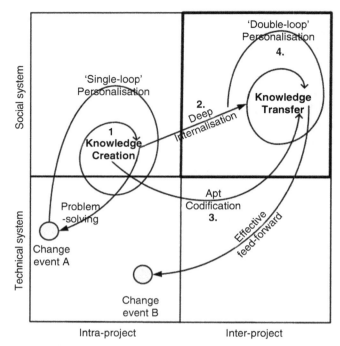

Figure 6.6   Knowledge-based reactive change process.

social networks will create the *'double-loop' personalisation*. This will, in turn, help to effectively *feed forward* knowledge created from past change events to future events and thereby achieve effective project change management. Hence, the key to a knowledge-based reactive change process lies in shifting the emphasis from the *intra-project technical system* to the *inter-project social system*.

6.6   Implications for Practice

Based on the knowledge-based theoretical model (see Figure 6.6), this section formulates guidelines to facilitate effective change management in construction projects. The guidelines are discussed under four headings: intra-project knowledge creation; deep internalisation; apt codification; and double-loop personalisation.

(1)   Intra-project knowledge creation

(a)   *Effectively identify and appropriately use team members' expertise in change problem- solving*

The extent to which team members hold past knowledge and the extent to which they know one another prior to discussions can improve the reactive change process. Therefore, team members need to know who knows what through prior interactions and make effective use of their knowledge during the reactive change process. For example, partnering arrangements can create prior relationships and also enable the use of specialist knowledge of sub-contractors where relevant.

(b)   Consider face-to face settings for change problem- solving

Tacit knowledge that originates at a collective level is the most strategically significant type of knowledge for effective project change management. Therefore, it is important to promote team interactions to create this tacit-collective knowledge. Virtual communication techniques are insufficient to this end, and face-to-face settings at regular intervals are crucial. For example, project teams can effectively use project progress meetings and special meetings where relevant to discuss change events.

These team discussions need to be effectively used to externalise team members' tacit knowledge. For example, by using visualisation techniques such as pictures, diagrams, sketches and models, and also by using examples from previous projects, team members can effectively share their tacit knowledge with others.

Teams can also be introduced to other socialisation activities to improve their tacit knowledge. For example, apprenticeships and training; participation in outside activities such as site visits and social events; and effective workplace arrangements such as open-plan layouts can assist to this end.

(c)   Create a knowledge-sharing team culture

It is important to create a team culture that uses shared language/narratives and to improve personal relationships through trust, care and openness. This can be facilitated by collaborative team approaches such as partnering arrangements and 'design-and-build' approaches. However, using collaborative arrangements does not automatically create knowledge-sharing cultures. Therefore, to promote knowledge sharing, further steps need to taken. For example, incentives (monetary or otherwise) can be introduced for good practices.

(2)   Deep internalisation

(a)   Encourage team members to learn and reflect after the change experience

Team members should be provided with time and resources to reflect, learn and experiment from their immediate change experience. For example, team members can hold post-project reviews by involving all key members in opportunities to discuss, share and reflect on lessons learnt. Project clients can encourage this activity by allowing time at project completion, and top management at multiple organisations can encourage this by giving sufficient time and motivation for such reflection.

(3)   Apt codification

(a)   Encourage team members to effectively codify their change experience

It is important to organise ideas raised in discussions and make conclusions. These need to be codified in documents such as review reports, manuals and databases. It is important to codify detailed information such as solutions considered and lessons learnt, rather than just a brief record of the final decision. Further, it is imperative to identify which party should maintain such documentation on behalf of a project and to make sure such documentation reaches other key members.

(b) Disseminate codified knowledge effectively within multiple organisations

The codified knowledge needs to reach individuals in the organisation. First, individuals need to be given effective access to information technology, which is a good medium for disseminating codified knowledge. For example, individuals can be given e-mail facilities and access to intranet and internet. Second, members should be made aware where knowledge resides within themselves. For instance, documents such as project files, databases and procedure manuals can be linked to knowledge maps that direct individuals to where a specific type of knowledge resides in the organisation.

(4)   Double-loop personalisation

(a)   Encourage face-to-face settings at multiple organisations

Team members should be allowed to interact with other members within their individual organisations, so that their experience can be shared through further socialisation. For example, regular meetings, brainstorming sessions and events such as conferences and social activities could be arranged to enhance socialisation.

Appendix A: Interview Guidelines

The interview questions are based on a selected change event and will be extended to general project data where appropriate.

This interview is structured in four sections. Following are the points covered by each section:

(1) Identify the background information of the selected change event.
(2) Explore the team interaction process during this change event.
   (Sections 1.0 and 2.0 will meet objective one of the study)
(3) Explore the variables that affect this change event.
   (Section 3.0 will meet objective two of the study)
(4) Identify the role of knowledge in this change event. Specifically how knowledge is captured, converted and re-used by the project stakeholders.
   (Section 4.0 will meet objectives three and four of the study)

The study expects to collect data from key stakeholders relevant to the selected change event (Contractor, Architect, Sub-contractor, Structural Engineer, Client Agent and the Client). The stakeholders will be interviewed based on the following guidelines and will be further provided with a brief 'closed questionnaire' in an attempt to collect both qualitative and quantitative data. Further the relevant documentation will be analysed.

The collected information will remain confidential. The interview transcripts will be sent back to the interviewees for their review and acceptance.

Project:
Interviewee:
Date:
Change event description:

1.0   Change situation identification

1. What was the evidence of this change?
2. What were the causes of this change?

Managing Change in Construction Projects: A Knowledge-Based Approach, First Edition.
Sepani Senaratne and Martin Sexton.
© 2011 by Sepani Senaratne and Martin Sexton. Published 2011 by Blackwell Publishing Ltd.

3. At what stage was it identified? How long it lasted?
4. How do you view this change (in terms of project/your firm)?
5. How did this change affect you and the other functions/parties?

2.0   Managing change process

6. Was there a formal change management process in place? Explain.
7. Who identified, evaluated, decided and approved the change?
8. How did the project team make decisions for change events (frequency, length, participants, style)?
9. Were there special meetings/discussions, workshops, brainstorming sessions and the like for evaluating and deciding the change? Explain.
10. How do you explain the availability of resources (including information/knowledge) in solving this problem (time to reflect, skills and expertise, physical and electronic resources)?
11. What technologies did the team use in sharing information/ communication (eg: e-mail, t/phone, networks, fax)?

3.0   Context and variables

Change-specific characteristics

12. How significant was this change (project time/cost/quality wise)?
13. How do you describe the level of complexity in this change situation (decision-making, technical)?
14. How do you describe the level of uncertainty in this change situation (programme, etc.)?

Group characteristics

15. Which project members actively participated for this change process?
16. What was your role in this change?
17. Explain the role of the facilitator/leader and the role of the client in this change decision-making.
18. Did the team share a common language? If not how did the differences in perception affect in reaching a decision?
19. How was the team geographically spread? Was there any effect on change?
20. Were the team members consistent throughout change? Was there any effect on change (members left, supply chain extended, replacements, same individual represented)?
21. Describe your working experience in the construction industry. How long? How many similar projects/period?
22. Explain your previous experience with this kind of change.
23. Explain your previous working relationships with the project team members and the client.

24. How do you explain the level of care that existed between parties in solving this problem (trustworthy, dependable, reliable, empathetic, helpful, lenient in judgement)?
25. How do you explain the behavioural norms that existed between the parties (sharing information openly and freely, open to criticisms, degree of consensus, togetherness, loyalty, value of diversity, tolerate failures)?
26. Explain how you felt as a team member (sense of identification)?

Project environmental characteristics

27. How did the corporate-level factors such as management structure and culture affect this change?
28. How do you describe the top management support during the change?
29. Were the following conditions present in your organisation both at firm and project level:
    vision of organisation in creating new knowledge;
    freedom/autonomy given to absorb knowledge;
    exposure to environmental fluctuations that led to breakdown of routines;
    redundancy of information – availability of common knowledge;
    requisite variety – availability of variety of information instantly?
30. Describe how your company values development of new ideas/concepts/best practices. Any incentives/rewards given?
31. Explain the opportunity to interact with other individuals in your organisation (e.g. personal networks, meetings)?

4.0   Knowledge capture, conversion and re-use

Knowledge capture

32. How did you prepare for team discussions? What information was collected? How (e.g. t/phone conversations, presentations, dialogues, etc. – verbal and bodily experience)?
33. What sort of feedback did you receive from prior direct interactions with client/end user?
34. What sort of feedback did you receive from prior direct interactions with suppliers/sub-contractors?
35. How did you share what you collected with the team (internal/external/prior team discussions)?

Socialisation

36. Did you learn anything by observing/being with the other team members?

37. Was there capacity for any member to train, observe or imitate seniors through this change experience?
38. Were there any training sessions/courses that influenced you in this issue?
39. Were there other activities that you participated in outside the organisation influential to you (e.g. site visit, fairs, social function)?
40. Were there any informal working patterns that existed during the change event (e.g. seated closely, community of practices, informal networks)?

Externalisation

41. Explain how you shared your previous experiences with others, during team discussions (example projects, myths, success and failure stories).
42. When others did not understand you how did you try to explain your ideas/abstract concepts/knowledge (use of examples, metaphors, analogies, use of language/terms)?
43. How did you convert your unorganised thoughts into concrete ideas for this change decision (models, pictures, formats, structures)?
44. Explain how you listened to others' opinions and thoughts.
45. How did you encourage others to express/clarify their thoughts?
46. Were any new concepts, thoughts and ideas brought to this issue by combining project team experiences/discussions?

Combination

47. Did you organise ideas raised in discussions and make conclusions/summaries?
48. What documents/reports did you refer to in reaching decisions (past project reviews)?
49. To what extent did you collect new documents and make connections with the old documents to work up new ideas/solutions?
50. What were the documents produced by incorporating knowledge gained by this change event (manuals, databases, publications, computer programmes)?
51. To what extent was the information/knowledge of this change event disseminated within other parts of the organisation?

Internalisation

52. After hearing a new idea or concept how did you compare it with your experience to help you comprehend the meaning?
53. Explain how you made sure that you understood others' speech/thoughts during dialogues.

54. Did you carry out experiments/trial-and-error work/exercises to improve your learning on the issues considered?

Knowledge re-use

55. Overall, what did you learn new by going through this change event?
56. How will this learning help you in managing this kind of change in future projects?
57. In case of parallel projects were knowledge and experience exchanged?
58. How was this change recorded and documented? By whom (project/ firm)?
59. To what extent was project feedback carried out after completion?

Appendix B: Example of an Interview Transcript

Interview Transcript – Client Agent's Perspective

Interviewee: X (actual names are changed to maintain confidentiality)

Date: 03rd June xxxx

Project: A

Change event description: *Flooring change*

1 **What was the evidence of this change?**
   At X (client) there is stage 1, estimating stage, in which stage they inform Y (client agent) that they need some renovations to the store. At stage 2, the design solutions are explored with the Architect. So this is the stage that the need to replace the store floor with terrazzo was identified. Therefore the original proposal was to replace the floor with terrazzo. This turned to a change case, when Z (D&B contractor) was appointed as the D&B contractor.

2 **What were the causes of this change?**
   At early stages of construction, Z informed X that terrazzo replacement requires lot more time. When Z pointed out to X that they require 6 weeks store closure for the terrazzo replacement with the cost implications, X was OK with the cost. But somebody within X found that they have not allocated the loss of trade for 6 weeks.
   Normally what happens in X store refurbishment is to carry on the work while the store is in operation and at the final stage to have a 2 weeks shutdown to ready the store for public. But for terrazzo replacement it requires 6 weeks and this had not been allowed in the overall figures of X (human error).

3 **At what stage was it identified? How long did it last?**
   Started with the letter dated 25th of February of Z sending the programme and the different solutions to X, till around 23rd of May and

Managing Change in Construction Projects: A Knowledge-Based Approach, First Edition.
Sepani Senaratne and Martin Sexton.
© 2011 by Sepani Senaratne and Martin Sexton. Published 2011 by Blackwell Publishing Ltd.

some more time thereafter. The issue went on for this long because X wanted a terrazzo floor and they did not want to allow a complete shutdown of 6 weeks nor they did not want to change the completion date, which led the team to find different solutions for the problem.

4   **How do you view this change (in terms of project/your firm)?**
From project point of view, initially it was a major problem; where we have an important client who is not getting what he wants. It is always difficult to explain the client, as it was his mistake in the first place. By the end of the project we made a happy client with a satisfactory solution (tile floor). Therefore the change was beneficial, because people were more than happy with the new floor, no defects were found after completion and there were savings by not going for the terrazzo floor.
    In terms of project team tasks, it was disruptive. It involved a waste of management time. Due to the delay of X in making a decision, the team was discussing the issue at several progress meetings and working around solutions for months. So there was a lot of wasted time.

5   **How did this change affect you and the other functions/parties?**
Mainly the contractor's functions were affected. The Architect was involved at early stages of design solutions with Z, but later on when the specialist flooring sub-contractor (Q) was invited to the discussion stage, the Architect's work was not affected much. Most solutions were produced as reports rather than design/sketches. So no drawing revisions involved. Flooring sub-contractor's work was affected, as he could not plan the work until a decision was reached. Other sub-contractors such as drainage sc were affected too. This work did not affect the other construction work, as the decision was somehow made before the scheduled time for flooring, in the work programme.

6   **Was there a formal change management process in place? Explain.**
Yes this is the CRF procedure that is practised by X. The procedure is when there is a change from the original plan, the contractor has to first submit to the client agent the cost and time implications of the proposed change. We check and agree on this. Then give our recommendations to the client in a CRF form and when the client approves this form by signing, only then does the change implementation take place. This is the common procedure for any client. The forms are maintained by QS (client's agents – Y).

7   **Who identified, evaluated, decided and approved the change?**
Contractor identified the change. Y (our QS) with Z evaluated. The client gave the final decision and approval.

8   **How did the project team make decisions for change events (frequency, length, participants, style)?**
For the evaluation and decision-making of change case, the specialist sc (Q)'s input was used. We discussed this in team meetings.

Initially we tried to explore the scope of the problem. We first did a survey of the screed by tapping the floor to identify hollow areas and so on. If we did the original floor design (replacing of terrazzo floor) this testing is not required. The survey was specific to the flooring problem because we tried to find solutions other than complete replacement of the floor. We knew that the existing tiles of the floor are lifting in some areas because of the bad screed and to identify the total scope of this problem this survey was carried out. So Z came up with solutions with Q after these surveys.

These solutions were discussed at design team meetings and also in special meetings.

DTMs are held once every two weeks, and last 2–3 hours. Regular participants were Z, X rep, Architect, Structural Engineer, Services Engineers and Y. SC's participation very rare. They are involved only when there's a specific problem. Q participated at one DTM.

The client representative was inconsistent throughout and that was one problem for delaying the decision. Everybody talks but no one decides. For some DTMs, the store delivery manager (end user) participated too.

The decision-making was by Client party. It was a mixture or a team decision-making process that involved the store manager, H/O rep, etc.

9   Were there special meetings/discussions, workshops, brainstorming sessions and the like for evaluating and deciding the change? Explain.
A special meeting was organised by Y for which a client's rep was also invited. There were special meetings between Z and Q as well.

There were no other special forums other than this special meeting. This special meeting was sort of a brainstorming session, in that we discussed the options in detail, is this the right solution in terms of cost and time and client's preferences, etc.?

10   How do you explain the availability of resources (including information/knowledge) in solving this problem?
Physical and electronic resources were not a problem. We also received skills and expertise from Q and Z. There was a time constraint as the completion date was fixed. It was not a resource problem rather a decision-making problem by the client.

11   What technologies did the team use in sharing information/communication?
E-mail mainly. Also telephones and fax. The use of information channel as a means of day-to-day communication was limited. Rather it was a shared database, which provided project documentation (not all). For the flooring issue information channel did not help that much.

12   How significant was this change (project time/cost/quality wise)?
Very significant.
Time – it involved 4–6 weeks' construction time (*no effect on time as the end date was fixed*).

Cost – original terrazzo floor costed £450,000 (of £7 million project) therefore was a substantial item (*the final solution was to over-tile and this costed about £250,000, therefore was a saving*).

Quality – the final solution was good quality wise, though we had a long list of defects in other areas there was no defects found with respect to the floor.

13   How do you describe the level of complexity in this change situation?
Quite simple technically as it required over-tiling and this was not a problem with a specialist sub-contractor.

Process was simple again as it is about identifying the problem and coming up with a solution by evaluating options. This was made difficult by the individuals involved doing their tasks slowly. But the actual change itself was simple.

14   How do you describe the level of uncertainty in this change situation (programme, etc.)?
Very high. The issue went on without a decision for a long period. The closer the end date, the more uncertain it became.

The work programme was not affected. But the late decision made the contractors wait for a long period to plan/organise their work. Client was made aware if he did not make a final decision by a given time, the end date needs to be extended. So, though they were late, they managed not to exceed this time limit.

15   Which project members actively participated for this change process?
Z, Y, Q, X were the active participants; Architect not greatly; Structural Engineer not very much.

16   What was your role in this change?
Our role was to agree on the cost and the programme implications and get the change order signed by the client and then instruct the contractor. Any change needs to go through client approval.

17   Explain the role of the facilitator/leader and the role of the client in this change decision-making.
Facilitator – In meetings between client and us, a person from X H/O facilitated the meetings.

Z facilitated the DTM and we led the special meeting.

Client's role was to make the final decision. Initially the client made the situation difficult by requesting a quality floor (terrazzo) without compromising the time. At this time all other X stores were decided to be terrazzo floor (trend) and therefore it was an internal problem for them to sort out.

18   Did the team share a common language? If not how did the differences in perception affect reaching a decision?
There was no problem within the team. One instance where a difficulty arose was when Q suggested a specific tiling, which they had used

before and was known to them. This was difficult to put across to the client, as they had no idea of this tiling type. Use of technical terms between the team was not a problem.

But it was always a problem when communicating with the client and making him understand the issues, especially making the client aware that he cannot have time, cost and quality at the same time. The main people within X wanted a terrazzo floor, there were specific project people who preferred other floor types. So there was confusion of preferences within the client team as well.

19   **How was the team geographically spread? Was there any effect on change?**
P (Architect), Z & Y are based in Bristol. The client at Bristol & H/O team at London. Q are at Kent and others were spread at different locations. But the key people were close enough and this did not affect this change very much. Even Q participated for a few meetings with regard to flooring issue. So no big effect from this.

20   **Were the team members consistent throughout change? Was there any effect on change?**
No effect from team consistency. All members who were involved in this issue were same. The only inconsistency was the changes in client representatives, which delayed their decision.

This is how X operates. They have the store manager, who participates in meetings (he doesn't have any authority although he knows what X wants); the store delivery manager who is the liaison between the store and the X head office (his involvement in discussions is minimum); and a key person with authority (he does not come to the site and is not updated on project issues). That is the problem with regard to their structure.

21   **Describe your working experience in the construction industry. How long? How many similar projects/period?**
Since 1974, I have been working in the construction industry. I have experience in retail projects but this is my first experience in X retail project.

That is as an individual; as a practice (Y) we have been working with X for 20 years. There is a team (of 5 people from total of 80) at Y who work in X projects. I got involved because this team was very busy with other X projects during this time.

22   **Explain your previous experience with this kind of change.**
Not this specific flooring problem. I had experience with floor problems like cracking, breaking-up, scaling but not floor tile lifting like in this case.

23   **Explain your previous working relationships with the project team members and the client.**
Personally, this client is the first time for me, but with Y 20 years of working relationship.

Personally and as a practice, worked with Z for 25 years.

With P as a practice, we had work with them on other projects as well (e.g. post office) and personally this is my second project with them. And with some other members as well (e.g. civil engineer).

24 How do you explain the level of care that existed between parties in solving this problem?

All the team members were very helpful. All wanted to solve the problem.

Trustworthy, yes. But there was a small incident between client and Q. When the client agreed to Q's suggestion, Q later changed their minds saying that the agreed option was not workable and introduced new solutions. This slightly led the client to reduce their confidence in Q. But other than that everyone was trustworthy and helpful.

25 How do you explain the behavioural norms that existed between the parties?

No problem with sharing information openly and freely. No problem with any others (openness to criticisms, degree of consensus, togetherness, loyalty) as well.

Value of diversity: the rest of the team did not have the expertise; Q's specialist knowledge was therefore valued and welcomed.

Tolerating failures: X is not very keen on this.

26 Explain how you felt as a team member?

No problem. The team accepts me/Y as the client agent. More than willing to do what we ask them to do.

27 How did the corporate-level factors such as management structure and culture affect this change?

Internally the structure was: I was doing the Client agent's (the management) role and there was another member from Y working as the quantity surveyor for this job. We split the work between us and had no problem with this structure.

No cultural problems, as we have been working with X for years and are quite used to them.

28 How do you describe the top management support during the change?

Internally there's a partner responsible for X projects. Because this issue went on for some time, we brought this matter to his attention. We normally have once a month meeting between Y and X, where all X projects are subjected to discussion. This was discussed at this high level at one of these meetings. So no problem with the top management support.

29 Were the following conditions present in your organisation both at firm and project level?

Vision of organisation in creating new knowledge: Y are always keen on new ideas particularly profitable ones, e.g. new surveys, etc.

Freedom/autonomy given to absorb knowledge: they are very good on that. We have a training academy, and there's Y internet accessible to all employees. In that all training courses' (for all levels from school leavers to partners) details are given.

Exposure to environmental fluctuations that led to breakdown of routines: not sure about this, normally there's no breaking of routine tasks.

Redundancy of information – availability of common knowledge: Y internet provides us with this common knowledge. This includes procedures, guides and checklists for various tasks. There is a common knowledge but not specific knowledge such as how to deal with a floor tile lifting issue. If we put these details the website will be so wielding, unless these are put under managerial issues.

Requisite variety – availability of a variety of information instantly: there is a pool of information available instantly through this website. For example, if I have a site visit on practical completion on Monday morning, I will access the internet and get a checklist in preparation.

30 **Describe how your company values development of new ideas/concepts/best practices. Any incentives/rewards given?**
Y are very keen on new ideas/concepts if they are new ways of making money. They look for best practices to keep client happy. Rewards are not specific. But with a new idea if you generate money and work for the company you are rewarded/recognised.

31 **Explain the opportunity to interact with other individuals in your organisation (e.g. personal networks, meetings).**
We have a once a month meeting with all Y's employees. This is a forum where all employees report what they are doing in general, but not details as to specific problems (e.g. flooring).

We also have e-mail facilities through Y's website. We tend to use e-mails among the staff often (avg. 1/4 from Y staff). It is common to send an e-mail, asking certain questions, to others. So we share knowledge this way too.

We moved from another office premises two years ago to this new office, which is an open-plan layout. This provides more interactions with the staff.

Each year we organise a conference where all members attend and where bosses speak about our work.

As individuals we are members of different external organisations, committees, etc. As per our job, the more that we get to know the better.

32 **How did you prepare for team discussions? What information was collected? How?**
Normally a few days before scheduled meetings, I ring Z and make sure whether they have answers for the client and what issues they are gong to raise at the meeting, etc., so that we know where things are coming from.

Yes, we use telephones, e-mails. In the specific meeting that we had Q made a sort of a presentation that explained the different options (not with ppt).

33 **What sort of feedback did you receive from prior direct interactions with client/end user?**
One bodily experience gained from the client was understanding his unhappiness when Q changed their mind on an agreed solution.

34 **What sort of feedback did you receive from prior direct interactions with suppliers/sub-contractors?**
Since we did not deal directly with sub-contractors and suppliers there was no such feedback received.

35 **How did you share what you collected with the team (internal/external) prior team discussions?**
Internally I share the information with the QS and in some instances with the Y Partner who is responsible for X projects.
Externally we shared information prior to discussion with Z.

36 **Did you learn anything by observing/being with the other team members?**
I have not noticed any learning experience as such.

37 **Was there capacity for any member to train, observe or imitate seniors through this change experience?**
At project meetings I had opportunity to observe other members and their work.
Within our firm QS comes along for all discussions and not sure whether he gained experience by observing/imitating others/me.
Normally when a new project is commenced a senior introduces me to the project by a site visit but thereafter I work on the project on my own. But in the past we had a system of accompanying junior members for site visits and meetings. That is no more.

38 **Were there any training sessions/courses that influenced you in this issue?**
No, not for this kind of issues.

39 **Were other activities that you participated in outside the organisation influential for you?**
I do not tend to do these much. We had site visits for this project. Fairs, no.
We do participate in social activities, either organised by other organisations such as RICS, Faculty of Building, Contractors or by Y, that invite clients. But these are networking rather than sharing technical experience.

40 **Were there any informal working patterns that existed during the change event?**
No.

41 **Explain how you shared your previous experiences with others, during team discussions.**

I have not come across this problem before, nor had any other key members. That is why we brought Q in. In general, we bring previous experience to issues but not for this specific one.

One such instance was, last week we made a site visit with a client for a new project (demolish the existing and reconstruct a new marketing building). So when I looked at it I proposed that the client needs an asbestos ceiling survey for this and I further told him a specific type, a fully intrusive one. This came from my previous experience on working with asbestos ceilings.

42 **When others did not understand you how did you try to explain your ideas/abstract concepts/ knowledge?**

Others tend to understand all what I say. They rarely say they don't. If they don't, I will repeat it or try to explain it in a different way.

I do not have a habit of using metaphors, analogies, etc. to express my personal knowledge. I like to keep it as simple as possible and express it in few words.

43 **How did you convert your unorganised thoughts into concrete ideas for this change decision?**

The team put down the unorganised thoughts into explicit forms for this flooring issue by developing a programme (like a flow chart).

I personally do not have a habit of using things like models, pictures, etc. to help explaining in discussions.

44 **Explain how you listened to others' opinions and thoughts.**

Generally at meetings when we question a person and when the other replies, we carefully listen to him and cross-question; ask them is that reasonable.

Especially when dealing with builders, as client's agents we need to question them about the accuracy of information they put forward at the discussions.

45 **How did you encourage others to express/clarify their thoughts?**

Always make sure that everybody has a chance to put their views across. Some members are often quiet and just keep listening to what is going on. I encourage them to express in meetings. Otherwise we will not get the best solution for the problem in hand.

46 **Were any new concepts, thoughts and ideas brought to this issue by combining project team experiences/discussions?**

Not really, as I said this is a new issue that the team did not experience before.

47 **Did you organise ideas raised in discussions and make conclusions/ summaries?**

Yes. First I make sure that everyone understands what the problem is and seek ideas from the team on alternative ways of solving the problem.

Some members have preferred solutions and they always speak about that particular one. So before coming up with conclusions, I make sure that everyone has expressed their views. Then I draw a summary and read this to the team. For example, Solution 1 is this and it leads to these implications, etc. This way all members' ideas are considered and summarised. Finally, this is forwarded to the client with our recommendations on the best one.

48   **What documents/reports did you refer to in reaching decisions?**
We referred to Q's reports. We referred to others' work rather than building solutions by ourselves (that is our role).

49   **To what extent did you collect new documents and make connections with the old documents to work up new ideas/solutions?**
Not really. We only referred to Q's reports.

50   **What were the documents produced by incorporating knowledge gained by this change event?**
No. This is the problem. Unfortunately, it is only the individuals involved who carry this knowledge in their heads and not passed on to others by making them explicit.

51   **To what extent is the information/knowledge of this change event disseminated within other parts of the organisation?**
Again no. In case of other project specific issues too this problem exists.

52   **After hearing a new idea or concept how did you compare it with your experience to help you comprehend the meaning?**
Yes. That happens. For example we checked the programmes submitted by Z. Since I've have experience with reading programmes, I was able to decide that the programmes were reasonable.
   Since our job is to check builders' work, this always happens. We check their work and combine this with our experience to say whether it is OK or not.

53   **Explain how you made sure that you understood others' speech/ thoughts during dialogues.**
When I did not understand them I asked them the same question in a different way or asked for clarification. If they use technical jargon, I may ask them to detail or explain in simple way.

54   **Did you carry out experiments/trial-and-error work/exercises to improve your learning on the issues considered?**
On this specific problem we made screed tests to decide on solutions. In general, not really. Probably this could happen when builders/clients want to do it in a different or a new way. But not that much.

55 **Overall, what did you learn new by going through this change event?**
I learnt how to deal with the client (how to make sure he understands and how long he needs to understand, etc.).

Also I learnt various possible solutions that can be considered when we face a floor tile-lifting problem. And practical things like the time that it takes for this kind of floor replacement and some rule of thumb methods used for testing the floor.

(Z did the work in night shifts. They had to use aisles to make it dust free. Not sure whether this is normal to X or if they used aisles for the first time.)

56 **How will this learning help you in managing this kind of change in future projects?**
Well, when the contractor or the consultants bring solutions to the problem, if they have not considered the options that we found in this case, I would ask what about this alternative?

Generally the process could be made efficient and effective next time, only if the client makes prompt decisions. So as you see this depends on individuals involved.

57 **In case of parallel projects were knowledge and experience exchanged?**
No. I was working with other projects simultaneously. But not on retail projects apart from another X extension work. In this X one there was no work to the existing store and the extension involved a new floor. So had no chance to exchange this flooring knowledge.

This is a very specific problem, so not often you experience two instances by yourself.

58 **How was this change recorded and documented? By whom (project/ firm)?**
On CRF system. These are filed and not often referred back.

If we recall a specific thing that relates, we will go back and refer to this in project files.

There're so many project files in the firm and other members simply will not know where to look, except in project files that they worked on by themselves.

What seems to happen in projects there's always particular problems and the individuals involved go through the experience and learn the best solution for the given problem. But this does not pass to other members. This same flooring problem may be experienced at the same time by a colleague, without my knowledge. We do not have a way of making sure that everybody knows what everybody else is doing.

59 **To what extent was project feedback carried out after completion?**
For X projects this is done by way of project review meetings. They have a standard timing for these meetings. First one 3 weeks after the handing over date, next at 3 months, 11 months and 12 months.

X maintains a KPI system and they review the performance of the consultants and contractors based on performance criteria and by this only they decide who to select for their next job.

On another project we carried out a detailed project review after project completion, by request of the client. Quite often this is beneficial, but it depends on how freely and openly you are able to put things forward (e.g. criticising others' mistakes, accepting your mistakes, etc.).

Appendix C: Comparison Between Propositions and Codes

P ref.	Propositions description	N ref.	Free node description
	Contextual factors		
P1	The reactive change process is very complex and ill-structured rather than simple and well-structured, due to various contextual factors that could be identified under process characteristics, group characteristics, organisational characteristics and wider-environmental characteristics.		
	Process characteristics		
	design-driven triggers	24	Late identification of change
		26	Lengthy problem-solving process
		33	Omission due to human error
	role of client	9	Client dm focused on end users
		11	Client not compromising requirements
		14	Complexity from multi-headed client
		15	Difficulty due to slow client dm
	task dependency	7	Change, beneficial for the client
		8	Change, disruptive for the team
		13	Complexity due to change
		42	Rework due to change
		46	Significance of change impact
		49	Uncertainty due to change
		5	Availability of timely info for dm
	Group characteristics		
	team culture	34	Open K sharing culture
		43	Role of each team member
		12	Common language between team
		17	Good teamwork due to partnering
		23	Language depends on nature of issue
		27	Fewer newcomers due to partnering
	team communication	39	Processes in place to manage change
		52	Use of technology for communication

(Continued)

Managing Change in Construction Projects: A Knowledge-Based Approach, First Edition.
Sepani Senaratne and Martin Sexton.
© 2011 by Sepani Senaratne and Martin Sexton. Published 2011 by Blackwell Publishing Ltd.

Appendix C: *(cont'd)*

P ref.	Propositions description	N ref.	Free node description
	team consistency	10	Client involvement inconsistent
		44	Same team throughout the change
	team leadership	25	Leader depends on change situation
	Organisational characteristics		
	Structure and culture	18	Influence from higher org level
		35	Opportunity for KC at org level
		36	Org structure shaped by partnering
	Knowledge properties		
P2	The knowledge that the project team	2	An experienced team
	members use in the reactive change	28	Less opportunity to use explicit K
	process is:	41	Reliance-sharing of past experience
	more tacit than explicit;	51	Use of documents-info for dm
	more collective (mutual) than	31	Manage change as a team
	individual; and		
	more situational than prompted.	45	Scope for intra-project KC
	Knowledge identification and utilisation		
P3	Project team members, who know		
	where the knowledge resides within	37	Prior relationships with the team
	the team through prior interactions,		
	are better at identifying and actively	22	Key members in the change process
	utilising relevant team members'	53	Utilisation of specific K
	knowledge, during the reactive change		
	process, compared to project team		
	members who do not know where the		
	knowledge resides within the team.		
	Knowledge conversion		
P4	Project teams are more likely to create		
	new knowledge based on the existing		
	knowledge, during the reactive change		
	process, through a natural flow of		Support through sub-propositions
	knowledge rather than through a full		
	cycle of the knowledge conversion		
	process.		
SP4.1	Project teams who interact with other		
	team members regularly through	32	Need for face-to-face meetings
	face-to-face settings, during the reactive	41	Reliance-sharing of past experience
	change process, are better at utilising	47	Team interactions across distance
	existing tacit knowledge and creating	54	Utilising tacit K by socialisation
	new tacit knowledge, compared to		
	project teams who do not interact		
	regularly through face-to-face settings.		
SP4.2	Project teams who actively use		
	visualisation techniques during team	6	Change options, as new K
	discussions that arise during the	16	Encourage others to express thoughts
	reactive change process, are better at	55	Ways to externalise K at discussions
	expressing and externalising their tacit		
	knowledge, compared to project teams		
	who do not use such techniques.		

P ref.	Propositions description	N ref.	Free node description
SP4.3	Project teams unlikely to effectively combine and codify externalised tacit knowledge arising out of team discussions in reaction change situations.	1 29	Ad hoc codification done by each org Limited codification of new K
SP4.4	Project teams are more likely to acquire superficial learning through the change experience, during the reactive change process, rather than effective internalisation through reflection.	19 45 48	Internalising K gained via change Scope for intra-project KC Time intensity by parallel projects
	Inter-project knowledge transfer		
P5	The new knowledge that is created during the reactive change process is transferred to multiple organisations, for potential re-use in future projects, through personalisation strategies, rather than through codification strategies.		Support through sub-propositions
SP5.1	The new knowledge created during the reactive change process is re-used in future projects through the individuals involved during the process, rather than through the codified documents.	3 4 19 28	Applying K across projects Applying K in future projects Internalising K gained via change Less opportunity to use explicit K
SP5.2	The new knowledge created during the reactive change process is disseminated and made available to the wider organisation for potential future re-use, through interactive settings between the organisational members, rather than through effective dissemination of codified documents.	20 21 30 38 40 50	Inter-project KT via cyberspace Inter-project KT via face-to-face Limited K sharing at social events Problems in disseminating K Project feedback through meetings Unique nature of change

dm, decision-making; K, knowledge; KC, knowledge creation; org, organisation.

References

Adair, J. (1986) *Effective Teambuilding*. Gower Publishing.

Akinsola, A.O., Potts, K.F., Ndekugri, I. & Harris, F.C. (1997) Identification and evaluation of factors influencing variations on building projects. *International Journal of Project Management*, **15**: 263–267.

Akintoye, A. (1994) Design and Build; a survey of construction contractors' views. *Construction Management and Economics*, **12**: 155–163.

Al-Ghassani, A.M., Kamara, J.M., Anumba, C.J. & Carrillo, P. (2004) An innovative approach to identifying knowledge management problems. *Engineering, Construction and Architectural Management*, **11**: 349–357.

Alvesson, M. (2003) Beyond neopositivists, romantics and localists: a reflexive approach to interviews in organisational research. *Academy of Management Review*, **28**: 13–33.

Anumba, C.J. (2009) Towards next-generation knowledge management systems for construction sector organizations – Editorial. *Construction Innovation*, **9**: 245–249.

Anumba, C.J., Ugwu, O., Newnham, L. & Thorpe, A. (2001) A multi-agent system for distributed collaborative design. *Logistics Information Management*, **14**: 355–366.

Argyris, C. & Schon, D. (1978) *Organisational Learning*. Addison-Wesley, Reading, MA.

Atkinson, A.R. (1999) The role of human error in construction defects. *Structural Survey*, **17**: 231–236.

Augier, M. & Vendelo, M.T. (1999) Networks, cognition and management of tacit knowledge. *Journal of Knowledge Management*, **3**: 252–261.

Augier, M., Shariq, S.Z. & Vendelo, M.T. (2001) Understanding the context: its emergence, transformation and role of tacit knowledge. *Journal of Knowledge Management*, **5**: 125–136.

Austin, S., Newton, A., Steele, J. & Waskett, P. (2002) Modelling and managing project complexity. *International Journal of Project Management*, **20**: 191–198.

Baiden, B.K., Price, A.D.F. & Dainty, A.R.J. (2006) The extent of team integration within construction projects. *International Journal of Project Management*, **24**: 13–23.

Barlow, J. & Jashapara, A. (1998) Organisational learning and inter-firm 'partnering' in the UK construction industry. *The Learning Organisation*, **5**: 86–98.

Barrett, P. & Stanley, C. (1999) *Better Construction Briefing*. Blackwell Publishing Ltd., Oxford.

Managing Change in Construction Projects: A Knowledge-Based Approach, First Edition.
Sepani Senaratne and Martin Sexton.
© 2011 by Sepani Senaratne and Martin Sexton. Published 2011 by Blackwell Publishing Ltd.

Barrett, P.S. & Barrett, L.C. (2003) Research as a kaleidoscope on practice. *Construction Management and Economics*, 21: 755–766.

Bennett, R.H. (1998) The importance of tacit knowledge in strategic deliberations and decisions. *Management Decision*, 36: 589–597.

Bhatt, G.D. (2002) Information dynamics, learning and knowledge creation in organisations. *The Learning Organisation*, 7: 89–98.

Black, C., Akintoye, A. & Fitzgerald, E. (2000) An analysis of success factors and benefits of partnering in construction. *International Journal of Project Management*, 18: 423–434.

Blacker, F. (2002) Knowledge, knowledge work and organisations: an overview and interpretation. In *Strategic Management of Intellectual Capital and Organisational Knowledge* (eds C.W. Choo & N. Bontis). Oxford University Press, New York; pp. 47–62.

Bourgeois, L.J. (1981) On the measurement of organisational slack. *Academy of Management Review*, 6: 29–39.

Bower, D. (2000) A systematic approach to the evaluation of indirect costs of contract variations. *Construction Management and Economics*, 18: 263–268.

Bresnen, M. & Marshall, N. (2000) Partnering in construction: a critical review of issues, problems and dilemmas. *Construction Management and Economics,* 18: 229–237.

Bresnen, M., Edelman, L., Swan, J., Laurent, S., Scarbrough, H. & Newell, S. (2002) Cross-sector research on knowledge management practices for project-based learning. Paper presented at the *European Academy of Management 2nd Annual Conference*, Sweden.

Bresnen, M., Edelman, L., Newell, S., Scarbrough, H. & Swan, J. (2003) Social practices and the management of knowledge in project environments. *International Journal of Project Management*, 21: 157–166.

Burati, J.L., Farrington, J.J. & Ledbetter, W.B. (1992) Causes of quality deviation in design and construction. *Journal of Construction Engineering and Management*, 118: 34–49.

Burnes, B. (2000) *Managing Change: A Strategic Approach to Organisational Dynamics,* 3rd edn. Pearson Education.

Busseri, M. & Palmer, J. (2000) Improving teamwork: the effect of self-assessment on construction design teams. *Design Studies*, 21: 223–238.

Chan, P., Cooper, R., Carmichael, S., Tzortzopoulos, P., McDermott, P. & Khalfan, M. (2004) Does organisational learning create a learning environment? Conceptual challenges from a project perspective. Paper presented at the *20th Annual ARCOM Conference*, Herriot-Watt University, Edinburgh.

Chapman, R.J. (1999) The likelihood and impact of changes of key project personnel on the design process. *Construction Management and Economics*, 17: 99–106.

Cheung, S.O., Thomas, S., Lam, K.C. & Yue, W.M. (2001) A satisfying leadership behaviour model for design consultants. *International Journal of Project Management*, 19: 421–429.

Chinyio, E.A., Olomolaiye, P.O., Kometa, S.T. & Harris, F.C. (1998) A needs-based methodology for classifying construction clients and selecting contractors. *Construction Management and Economics*, 16: 91–98.

Chua, A. (2002) The influence of social interaction on knowledge creation. *Journal of Intellectual Capital*, 3: 375–392.

CII (1991) *Construction Changes and Change Orders; their Magnitude and Impact,* SD-66. Construction Industry Institute, Austin, TX.

CII (1994) *Project Change Management*. Special publication 43-1. Construction Industry Institute, Austin, TX.

CII (1997) *Modeling the Lessons Learnt Process*. RS123-1. Construction Industry Institute, Austin, TX.

CIRIA (2001) *Managing Project Change; A Best Practice Guide*. CIRIA C556, UK.

Cohen, W.M. & Levinthal, D.A. (1990) Absorptive capacity: A new perspective on learning and innovation. *Administrative Science Quarterly*, **35**: 128–153.

Collins, H.M. (1993) The structure of knowledge. *Social Research*, **60**: 96–116.

Constructing Excellence (2004) Effective Teamwork: A best practice guide for the construction industry. Available at: www.constructingexcellence.org.uk/pdf/document/Teamwork_Guide.pdf.

Cornick, T. & Mather, J. (1999) *Construction Project Teams: Making Them Work Profitably*. Thomas Telford.

Cox, A. & Townsend, M. (1997) Latham as half-way house: rational competence approach to better practice in construction procurement. *Engineering, Construction and Architectural Management*, **4**: 143–158.

Cox, I.D., Morris, J.P., Rogerson, J.H. & Jared, G.E. (1999) A quantitative study of post contract award design changes in construction. *Construction Management and Economics*, **17**: 427–439.

Cramton, C.D. (2001) The mutual knowledge problem and its consequences for dispersed collaboration. *Organisation Science*, **12**: 346–371.

Daft, R.L. & Lengal, R.H. (1986) Organisation information requirements, media richness and structural design. *Management Science*, **32**: 554–571.

Davenport, T.H. & Prusak, L. (1998) *Working Knowledge: How Organizations Manage What They Know*. Harvard Business School Press, Boston.

Denzin, N.K. & Lincoln, Y. (2000) *Collecting and Interpreting Qualitative Methods*, 2nd edn. Sage Publications.

Disterer, G. (2002) Management of project knowledge and experiences. *Journal of Knowledge Management*, **6**: 512–520.

Dixon, N. (2000) *Common Knowledge: How Companies Thrive by Sharing What They Know*. Harvard Business School Press.

Dubois, A. & Gadde, L. (2002) The construction industry as a loosely coupled system: implications for productivity and innovation. *Construction Management and Economics*, **20**: 621–631.

Easterby-Smith, M., Thorp, R. & Lowe, A. (1991) *Management Research: An Introduction*. Sage Publications, London.

Egan, J. (1998) *Rethinking Construction: Report of the Construction Industry Task Force*. HMSO, London.

Egbu, C., Kurul, E., Quintas, P., Hutchinso, V., Anumba, C.J. & Ruikar, K. (2003) *Knowledge Production, Resources & Capabilities in the Construction Industry – Work Package 1 – Final Report*. Available at: http://www.knowledgemanagement.uk.net/resources/kmfinalwp1.pdf.

Eisenhardt, K.M. (1989) Building theories from case study research. *Academy of Management Review*, **14**: 532–550.

Empson, L. (2001) Introduction: knowledge management in professional service firms. *Human Relations*, **54**: 811–817.

Fahey, L. & Prusak, L. (1998) The eleven deadliest sins of knowledge management. *California Management Review*, **40**: 265–276.

Fayek, A.R., Dissanayake, M. & Campero, O. (2003) *Measuring and Classifying Construction Field Rework: A Pilot Study*. Report presented to the Construction Owners Association of Alberta (COAA) Field Rework Committee.

Fiol, C.M. & Lyles, M.A. (1985) Organisational learning. *Academy of Management Review,* **10**: 803–813.

Fisher, S.G., Hunter, T.A. & Macrosson, W.D.K. (1997) Team or group? Managers' perceptions of the differences. *Journal of Managerial Psychology,* **12**: 232–242.

Fong, P.S.W. (2003) Knowledge creation in multidisciplinary project teams: an empirical study of the processes and their dynamic interrelationships. *International Journal of Project Management,* **21**: 479–486.

Fontana, A. & Frey, J.H. (2000) Interviewing – the art of science. In: *Collecting and Interpreting Qualitative Methods* (eds N.K. Denzin & Y. Lincoln). Sage Publications, pp. 361–376.

Franco, L.A., Cushman, M. & Rosehead, J. (2003) Project review and learning in the construction industry: Embedding a problem structuring method within a partnership context. *European Journal of Operational Research,* **152**: 586–601.

Galbraith, J.R. (1974) Organisation design: an information processing view. *Interfaces,* **4**: 28–36.

Gann, D.M. & Salter, A.J. (1998) Learning and innovation management. *International Journal of Innovation Management,* **2**: 431–454.

Gann, D.M. & Salter, A.J. (2000) Innovation in project-based, service enhanced firms: the construction of complex products and systems. *Research Policy,* **29**: 955–972.

Gieskes, J.F.B. & Broeke, A.M. (2000) Infrastructure under construction: continuous improvement and learning in projects. *Integrated Manufacturing Systems,* **11**: 188–198.

Gigone, D. & Hastie, R. (1993) The common knowledge effect: Information sharing and group judgement. *Journal of Personality and Social Psychology,* **65**: 959–974.

Gil, N., Tommelein, I.D., Kirkendall, R.L. & Ballard, G. (2001) Leveraging speciality-contractor knowledge in design-build organisations. *Engineering, Construction and Architectural Management,* **8**: 355–367.

Glaser, B.G. & Strauss, A.L. (1967) *Discovery of Grounded Theory: Strategies for Qualitative Research.*: Aldine, Chicago.

Gould, F.E. & Joyce, N.C. (2002) *Project Management in Construction,* 2nd edn. Prentice Hall, London.

Grant, R.M. (1996a) Toward a knowledge-based theory of the firm. *Strategic Management Journal,* **17**(s): 109–122.

Grant, R.M. (1996b) Proposing in dynamically-competitive environments: organisational capability as knowledge integration. *Organisation Science,* **7**: 375–387.

Grant, K.A. & Grant, C.T. (2008) Developing a model of next generation knowledge management. *Journal of Issues in Informing Science and Information Technology,* **5**: 571–590.

Grisham, T. & Walker, D.H.T. (2006) Nurturing a knowledge environment for international construction organizations through communities of practice. *Construction Innovation,* **6**: 217–231.

Gruenfeld, D.H., Mannix, E.A., Williams, K.Y. & Neale, M.A. (1996) Group composition and decision-making: How member familiarity and information distribution affect process and performance. *Organisational Behaviour and Human Decision Processes,* **67**: 1–15.

Gunasekaran, A. & Love, P.E.D. (1998) Concurrent engineering: a multi-disciplinary approach for construction. *Logistics Information Management,* **11**: 295–300.

Guzzo, R.A. & Salas, E. (1995) *Team Effectiveness and Decision-Making in Organisations.* Jossey-Bass Publishers.

Hanna, A.S., Jeffferey, S.R. & Vandenberg, P.J. (1999) The impact of change orders on mechanical construction labour efficiency. *Construction Management and Economics*, **17**: 721–730.

Hansen, M.T. (1999) The search-transfer problem: the role of weak ties in sharing knowledge across organisation subunits. *Administrative Science Quarterly*, **44**: 82–101.

Hansen, M.T., Nohria, N. & Tierney, T. (1999) What's your strategy for managing knowledge. *Harvard Business Review*, **77(s)**: 106–116.

Hobday, M. (2000) The project-based organisation: an ideal form for managing complex products and systems. *Research Policy*, **29**: 871–893.

Huber, G.P. (1991) Organisational learning: The contributing processes and the literatures. *Organisation Science*, **2**: 88–115.

Huber, P.H. (1996) Facilitating project team learning and contributions to organisational knowledge. *Creativity and Innovation Management*, **8**: 70–76.

Humphreys, P., Matthews, J. & Kumaraswamy, M. (2003) Pre-construction project partnering: from adversarial to collaborative relationships. *Supply Chain Management; An International Journal*, **8**: 166–178.

Ibbs, C.W., Kwak, Y.H., Ng, T. & Odabasi, A.M. (2003) Project delivery systems and project change: quantitative analysis. *Journal of Construction Engineering and Management*, **129**: 382–387.

Ingirige, M.J.B. (2004) A study of knowledge sharing in multinational construction alliances. Unpublished PhD thesis, University of Salford, Salford.

Jergeas, G.F. & Hartman, F.T. (1994) Contractors' construction-claims avoidance. *Journal of Construction Engineering and Management*, **120**: 553–560.

Jick, T.D. (1979) Mixing qualitative and quantitative methods: triangulation in action. *Administrative Science Quarterly*, **24**: 602–611.

Johnson, G. & Scholes, K. (1999) *Exploring Corporate Strategy*, 5th edn. Prentice Hall.

Kagioglou, M., Cooper, R., Aouad, G. & Sexton, M. (2000) Rethinking construction: the generic design and construction process protocol. *Engineering, Construction and Architectural Management*, **7**: 141–153.

Kamara, J.M., Anumba, C.J. & Evbuomwan, N.F.O. (2000) Developments in the implementation of concurrent engineering in construction. *International Journal of Computer Integrated Design and Construction*, **2**: 68–78.

Kamara, J.M., Augenbroe, G., Anumba, C.J. & Carrillo, P.M. (2002) Knowledge management in the architecture, engineering and construction industry. *Construction Innovation*, **2**: 53–67.

Karim, A. & Adeli, H. (1999) CONSCOM: An OO construction scheduling and change management systems. *Journal of Construction Engineering and Management*, 125(5): 368–376.

Katzenbach, J.R. & Smith, D.K. (1993a) *The Wisdom of Teams: Creating the High-Performance Organization*. Harvard Business School, Boston.

Katzenbach, J.R. & Smith, D.K. (1993b) The discipline of teams. *Harvard Business Review*, **Mar-Apr**: 111–120.

Kim, D.H. (1993) The link between individual and organisational learning. *Sloan Management Review*, **Fall**: 37–50.

Kogut, B. & Zander, U. (1992) Knowledge of the firm, combinative capabilities and the replication of technology. *Organisation Science*, **3**: 383–397.

Kolb, D.A. (1984) *Experiential Learning: Experience as the Source of Learning and Development*. Prentice-Hall, Englewood Cliffs, NJ.

von Krough, G. (1998) Care in knowledge creation. *California Management Review*, **40**: 133–153.

von Krogh, G., Ichijo, K. & Nonaka, I. (2000) *Enabling Knowledge Creation.* Oxford University Press, New York.

von Krogh, G., Nonaka, I. & Aben, M. (2001) Making the most of your company's knowledge: a strategic framework. *Long Range Planning*, **34**: 421–439.

Koskinen, K.U., Pihlanto, P. & Vanharanta, H. (2003) Tacit knowledge acquisition and sharing in a project work context. *International Journal of Project Management*, **21**: 281–290.

Kululanga, G.K., McCaffer, R., Price, A.D.F. & Edum-Fotwe, F. (1999) Learning mechanisms employed by construction contractors. *Journal of Construction Engineering and Management*, **125**: 215–223.

Kululanga, G.K. & McCaffer, R. (2001) Measuring knowledge management for construction organisations. *Engineering, Construction and Architectural Management*, **8**: 346–354.

Lam, A. (1997) Embedded firms, embedded knowledge: Problems of collaboration and knowledge transfer in global cooperative ventures. *Organisation Studies*, **18**: 973–996.

Latham, M. (1994) *Constructing the Team: Final Report of the Government/Industry Review of Procurement and Contractual Arrangements in the UK Construction Industry.* HMSO, London.

Lawson, B. (1997) *How Designers Think; The Design Process Demystified.* Architectural Press.

Lawson, M.B. (2001) In praise of slack: time is of the essence. *The Academy of Management Executive*, **15**: 125–135.

Leonard, D. & Sensiper, S. (1998) The role of tacit knowledge in group innovation. *California Management Review*, **40**: 112–132.

Leonard-Barton, D. (1995) *Wellsprings of Knowledge: Building and Sustaining the Sources of Innovation.* Harvard Business School Press. Boston.

Leung, M., Chong, A., Ng, T. & Cheng, M.C.K. (2004) Demystifying stakeholders' commitment and its impacts on construction projects. *Construction Management and Economics*, **22**: 701–712.

Levy, S.M. (2000) *Project Management in Construction,* 4th edn. McGraw-Hill, London.

Lewin, K. (1951) *Field Theory in Social Science.* Harper & Row, New York.

Li, H. & Love, P.E.D. (1998) Developing a theory of construction problem solving. *Construction Management and Economics*, **16**: 721–727.

Linde, C. (2001) Narrative and social tacit knowledge. *Journal of Knowledge Management*, **5**: 160–170.

Littelepage, G., Robinson, W. & Reddington, K. (1997) Effects of task experience and group experience on group performance, member ability, and recognition of expertise. *Organisational Behaviour and Human Decision Processes*, **69**: 133–147.

Loermans, F. (2002) Synergizing the learning organisation and knowledge management. *Journal of Knowledge Management*, **6**: 285–294.

Love, P.E.D. (2002) Influence of project type and procurement method on rework costs in building construction projects. *Journal of Construction Engineering and Management*, **128**: 18–29.

Love, P.E.D. & Li, H. (1998) From BPR to CPR – conceptualising re-engineering in construction. *Business Process Management Journal*, **4**: 291–305.

Love, P.E.D. & Li, H. (2000) Quantifying the causes and costs of rework in construction. *Construction Management and Economics*, **18**: 479–490.

Love, P.E.D., Gunasekaran, A. & Li, H. (1998) Concurrent engineering: a strategy for procuring construction projects. *International Journal of Project Management*, **16**: 375–383.

Love, P.E.D., Mandal, P. & Li, H. (1999) Determining the casual structure of rework influences in construction. *Construction Management and Economics*, **17**: 505–517.

Love, P.E.D., Mandal, P., Smith, J. & Li, H. (2000) Modelling the dynamics of design error induced rework in construction. *Construction Management and Economics*, **18**: 567–574.

Love, P.E.D., Holt, G.D. & Li, H. (2002) Triangulation in construction management research. *Engineering, Construction and Architectural Management*, **9**: 294–303.

Love, P.E.D., Irani, Z. & Edwards, D.J. (2004) A rework reduction model for construction projects. *Journal of IEEE Transactions on Engineering Management*, **51**: 426–440.

Lu, S. & Sexton, M. (2009) *Innovation in Small Professional Practices in the Built Environment*. Blackwell Publishing Ltd., Oxford.

Lurey, J.S. & Raisinghani, M. (2001) An empirical study of best practices in virtual teams. *Information & Management*, **38**: 523–544.

Manavazhi, M.R. (2004) Assessment of the propensity for revision in design projects through the dichotomous characterisation of design effort. *Construction Management and Economics*, **22**: 47–54.

Manavazhi, M.R. & Xunzhi, Z. (2001) Productivity oriented analysis of design revisions. *Construction Management and Economics*, **19**: 379–391.

Maqsood, T., Finegan, A. & Walker, D. (2006) Applying project histories and project learning through knowledge management in an Australian construction company. *The Learning Organization*, **13**: 80–95.

March, J.G. (1991) Exploration and exploitation organisational learning. *Organisation Science*, **2**: 71–87.

Maznevski, M.L. & Chudoba, K.M. (2000) Bridging space over time: global virtual team dynamics and effectiveness. *Organisation Science*, **11**: 473–492.

McCalman, J. & Paton, R.A. (2000) *Change Management*. Sage Publications, London.

McDermott, R. (1999) Why information technology inspired, but cannot deliver knowledge management. *California Management Review*, **41**: 103–117.

McElroy, M.W. (2000) Integrating complexity theory, knowledge management and organizational learning. *Journal of Knowledge Management*, **4**: 195–203.

Miles, M.B. & Huberman, A.M. (1994) *An Expanded Sourcebook – Qualitative Data Analysis*. Sage Publications.

Mintzberg, H. & Waters, J.A. (1982) Tracking strategy in an entrepreneurial firm. *Academy of Management Journal,* **25**: 465–499.

Mintzberg, H., Raisinghani, M. & Theoret, A. (1976) The structure of unstructured decision processes. *Administrative Science Quarterly*, **21**: 246–275.

Miozzo, M. & Ivory, C. (2000) Restructuring in the British construction industry: implications of recent changes in project management and technology. *Technology Analysis & Strategic Management*, **12**: 513–531.

Moore, D.R. & Dainty, A.R.J. (1999) Integrated project teams' performance in managing unexpected change events. *Team Performance Management*, **5**: 212–222.

Moore, D.R. & Dainty, A.R.J. (2001) Intra-team boundaries as inhibitors of performance improvement in UK design and build projects: a call for change. *Construction Management and Economics*, **19**: 559–562.

Muir, T. & Rance, B. (1995) *Collaborative Practice in the Built Environment*. E & F.N. Spon, London.

Nahapiet, J. & Ghoshal, S. (1998) Social capital, intellectual capital and the organisational advantage. *Academy of Management Review*, **23**: 242–266.

Nam, C.H. & Tatum, C.B. (1988) Major characteristics of constructed products and resulting limitations of construction technology. *Construction Management and Economics*, **6**: 133–148.

Nohria, N. & Gulati, R. (1996) Is slack good or bad for innovation? *Academy of Management Review*, **39**: 1245–1264.

Nonaka, I. (1994) A dynamic theory of organisational knowledge creation. *Organisation Science*, **5**: 14–37.

Nonaka, I. & Konno, N. (1998) The concept of 'ba': building a foundation for knowledge creation. *California Management Review*, **40**: 40–54.

Nonaka, I. & Takeuchi, H. (1995) *The Knowledge-Creating Company*. Oxford University Press, New York.

Nonaka, I., Byosiere, P., Borucki, C.C. & Konno, N. (1994) Organisational knowledge creation theory: a first comprehensive test. *International Business Review*, **3**: 337–351.

Nonaka, I., Toyama, R. & Konno, N. (2001) *SECI, Ba and Leadership: a Unified Model of Dynamic Knowledge Creation*. Sage Publications, London.

Pemberton, J.D. & Stonehouse, G.H. (2000) Organisational learning and knowledge assets – an essential partnership. *The Learning Organisation*, **7**: 184–193.

Perry, M. & Sanderson, D. (1998) Co-ordinating joint design work: the role of communication and artefacts. *Design Studies*, **19**: 273–288.

Polanyi, M. (1966) *The Tacit Dimension*. Harvard University Press, Boston.

Prencipe, A. & Tell, F. (2001) Inter-project learning: processes and outcomes of knowledge codification in project-based firms. *Research Policy*, **30**: 1373–1394.

Price, A.D.F. & Newson, E. (2003) Strategic management: consideration of paradoxes, processes and associated concepts as applied to construction. *Journal of Management Engineering*, **19**: 183–192.

Price, A.D.F., Bryman, A. & Dainty, A.R.J. (2004) Empowerment as a strategy for improving construction performance. *Leadership and Management in Engineering*, **Jan**: 27–36.

Reading Construction Forum (1995) *Trusting the Team, the Best Practice Guide to Partnering in Construction*. University of Reading.

Robinson, H.S., Carrillo, P.M., Anumba, C.J. & Al-Ghassani, A.M. (2002) Knowledge management for continuous improvement in project organisations. Paper presented at the *10th International Symposium on Construction Innovation and Global Competitiveness* – CIB W65, Cincinnati, Ohio.

Ruikar, K., Koskela, L. & Sexton, M. (2009) Communities of practice in construction case study organisations: questions and insights. *Construction Innovation*, **9**: 434–448.

Sampler, J.L. (1998) Redefining industry structure for the information age. *Strategic Management Journal*, **19**: 343–355.

Schindler, M. & Eppler, M.J. (2003) Harvesting project knowledge: a review of project learning methods and success factors. *International Journal of Project Management*, **21**: 219–228.

Schon, D.A. (1983) *The Reflective Practitioner: How Professionals Think in Action*. Basic Books, New York.

Scott, S. & Harris, R. (1998) A methodology for generating feedback in the construction industry. *The Learning Organisation*, **5**: 121–127.

Senaratne, S. & Hupuarachchi, A. (2009) Construction project teams and their development: case studies in Sri Lanka. *Architectural Engineering and Design Management*, **5**: 215–224.

Sexton, M. & Barrett, P.S. (2003) A literature synthesis of innovation in small construction firms: insights, ambiguities and questions. *Construction Management and Economics*, **21**: 613–622.

Sexton, M. & Barrett, P. (2004) The role of technology transfer in innovation within small construction firms. *Engineering, Construction and Architectural Management*, **11**: 342–348.

Simon, H.A. (1957) *Administrative Behaviour: A Study of Decision-making Processes in Administrative Organisation,* 3rd edn. Collier Macmillan Publishers, New York.

Simon, H.A. (1987) Making management decisions: the role of intuition and emotion. *Academy of Management Executive*, **Feb**: 57–64.

Simons, R.H. & Thompson, B.M. (1998) Strategic determinants: the context of managerial decision making. *Journal of Managerial Psychology*, **13**: 7–21.

Snowden, D. (2002) Complex acts of knowing: paradox and descriptive self-awareness. *Journal of Knowledge Management*, **6**: 100–111.

Sole, D. & Edmondson, A. (2002) Situated knowledge and learning in dispersed teams. *British Journal of Management*, **13(s)**: 17–34.

Sommerville, J. & Dalziel, D. (1998) Project teambuilding – the applicability of Belbin's team-role self-perception inventory. *International Journal of Project Management*, **16**: 165–171.

Spender, J.C. (1996) Making knowledge the basis of a dynamic theory of the firm. *Strategic Management Journal*, **17(s)**: 45–62.

Spender, J.C. & Grant, R.M. (1996) Knowledge and the firm: Overview. *Strategic Management Journal*, **17(s)**: 5–9.

Stasser, G., Stewart, D.D. & Wittenbaum, G.M. (1995) Expert roles and information exchange during discussion: The importance of knowing who knows what. *Journal of Experimental Social Psychology*, **31**: 244–265.

Sun, M. and Meng, X. (2009) Taxonomy for change causes and effects in construction projects. *International Journal of Project Management*, **27**: 560–572.

Szulanski, G. (2000) The process of knowledge transfer: a diachronic analysis of stickiness. *Organisational Behaviour and Human Decision Processes*, **82**: 9–27.

Teece, D.J. (1998) Capturing value from knowledge assets: the new economy, markets for know-how and intangible assets. *California Management Review*, **40**: 55–79.

Thomas, H.R. & Napolitan, C.L. (1995) Quantitative effects of construction changes on labour productivity. *Journal of Construction Engineering and Management*, **Sep**: 290–296.

Tombesi, P. (2000) Note: Modelling the dynamics of design error induced rework in construction: comment. *Construction Management and Economics*, **18**: 727–732.

Tuckman, B. (1965) Developmental sequence in small groups. *Psychological Bulletin*, **63**: 384–389.

Tushman, M.L. (1978) Technical communication in red laboratories: the impact of project work characteristics. *Academy of Management Journal*, **21**: 624–645.

Tushman, M.L. & Nadler, D.A. (1978) Information processing as an integrative concept in organisational design. *Academy of Management Review*, **3**: 613–624.

Udeaja, C.E., Kamara, J.M., Carrillo, P.M., Anumba, C.J., Bouchlaghem, N. & Tan, H.C. (2008) A web-based prototype for live capture and reuse of construction project knowledge. *Automation in Construction*, **17**: 839–851.

Van De Ven, A.H. & Poole, M.S. (1995) Explaining development and change in organisations. *Academy of Management Review*, **20**: 510–540.

Voropajev, V. (1998) Change management – a key integrative function of PM in transition economies. *International Journal of Project Management*, **16**: 15–19.

Walker, A. (2002) *Project Management in Construction,* 5th edn. Blackwell Publishing Ltd., Oxford.

Walker, D.H.T. (1998) The contribution of the client representative to the creation and maintenance of project inter-team relationships. *Engineering, Construction and Architectural Management*, **5**: 51–57.

Wang, C.L. & Ahmad, P.K. (2003) Organisational learning: a critical review. *The Learning Organisation*, **10**: 8–17.

Wantanakorn, D., Mawdesley, M.J. & Askew, W.H. (1999) Management errors in construction. *Engineering, Construction and Architectural Management*, **6**: 112–120.

Weick, K.E. (1995) *Sensemaking in Organisations*. Sage Publications, Thousand Oaks, CA.

Weick, K.E. & Quinn, R.E. (1999) Organisational change and development. *Annual Review of Psychology*, 361–381.

Winch, G. (2002) *Managing Construction Projects*. Blackwell Publishing Ltd., Oxford.

Wu, C., Hsieh, T., Cheng, W. & Lu, S. (2004) Grey relation analysis of causes for change orders in highway construction. *Construction Management and Economics*, **22**: 509–520.

Yang, H., Anumba, C.J., Kamara, J.M. & Carrillo, P. (2001) A fuzzy-based analytical approach to collaborative decision making for construction teams. *Logistics Information Management*, **14**: 344–354.

Yin, R. (1994) *Case Study Research: Design and Methods,* 2nd edn. Sage Publications, London.

Zhang, X., Sexton, M.G., Aouad, G., Goulding, J.S. & Kagioglou, M. (2004) A virtual prototyping system to aid hybrid concrete construction at the conceptual design stage. Paper presented at the *CIB World Congress: Building for the Future*, Toronto, Canada.

Index

absorption, 22, 28, 37, 39
absorptive capacity, 28, 38, 40
adaptation theories, 41
adversarial culture, 8–12, 124
Agyris, C., 41, 146
Anumba, C. J., 2, 11–12, 24, 123
architect, 55, 59, 61, 63, 65, 67, 71–3, 79,
    82–3, 85, 87–9, 91–3, 95, 97, 99,
    103–7, 111, 113
asset-based views, 23, 25, 44
autonomy, 35, 92, 120, 137

ba, 28, 34–6
Barrett, P. S., 4, 12, 14–16, 24, 37, 123
best practice guidelines, 3, 20
brainstorming, 72, 111, 132, 138, 140,
    143, 150
Bresnen, M., 8, 10–11, 19–20, 25–6,
    124, 133

capabilities
    combinative, 22
    core capabilities, 119, 121
    dynamic capabilities, 22
case study methodology, 47
change
    beneficial, 18, 20
    detrimental, 18, 20
    reactive, 5, 13, 16–17, 20, 28–9, 31, 36–7,
        40, 42–5, 48, 61, 63, 66–8, 75–7, 82–3,
        93–4, 98–9, 107–9, 113–14, 117,
        120–1, 123–7, 129, 131–3, 135–9, 141,
        143–9
    unexpected, 1
    unplanned, 4–5, 14–15, 42, 48, 118,
        122–3, 126, 128, 135, 137, 141

change causes, 17, 19, 21, 50, 108, 128
change context, 19–20, 36–7, 133
change effects, 1–3, 18, 21
change events, 4, 37, 42, 48, 51–3, 65, 67,
    76, 83, 88, 91–2, 97, 99, 101, 103, 108,
    114–15, 119, 122, 126, 131, 133, 136,
    138–9, 141, 144, 147–9
change experience, 36, 73, 75–6, 82,
    107–9, 111, 114, 126, 128, 131, 139,
    147, 149
change management, 2–5, 7, 12–13, 15,
    18–21, 24, 28, 50, 59, 62, 88–9,
    119–20, 135–6, 144, 147–9
change management guide, 21
change nature, 13, 62, 64, 93
change options, 55, 72, 98, 104, 127
change orders, 12, 14, 17
change process, 3, 7, 13, 15, 18, 21, 28, 34,
    41, 43–4, 55, 62, 66, 68, 73, 98–100,
    105, 108, 118, 120, 121, 127–8, 135–6,
    138, 140–1, 143–6
change record, 49, 51, 76, 105, 108, 127
change request forms, 73
change types
    anticipated, 14
    elective, 14, 21
    emergent, 13–15, 118
    operational, 14
    post-fixity change, 14
    required, 1, 2, 9, 14–15, 21, 26, 36, 40,
        53, 60, 67–8, 84, 113, 120, 137
    strategic, 10, 14
CII- Construction Industry Institute, 1, 3,
    13–14, 20, 25, 131
CIRIA, 1–3, 12, 14–15, 18, 20–1, 118
claims, 1, 14, 18, 40

Managing Change in Construction Projects: A Knowledge-Based Approach, First Edition.
Sepani Senaratne and Martin Sexton.
© 2011 by Sepani Senaratne and Martin Sexton. Published 2011 by Blackwell Publishing Ltd.